FETAL ENDOCRINOLOGY – AN EXPERIMENTAL APPROACH

Monographs in fetal physiology

Volume 1

General editor

P. W. Nathanielsz

Fetal endocrinology
An experimental approach

P.W. Nathanielsz, M.A., Ph.D., M.B.

Physiological Laboratory, Cambridge University, Cambridge, U.K.

1976

NORTH-HOLLAND PUBLISHING COMPANY
AMSTERDAM · NEW YORK · OXFORD

© Elsevier/North-Holland Biomedical Press, 1976

All rights reserved. No part of this publication may be reproduced, stored in a retrieval system, or transmitted, in any form or by any means, electronic, mechanical, photocopying, recording or otherwise, without the prior permission of the copyright owner.

North-Holland ISBN: 0 7204 0582 3

PUBLISHED BY:

Elsevier/North-Holland Biomedical Press,
335 Jan van Galenstraat, P.O. Box 211,
Amsterdam, The Netherlands

SOLE DISTRIBUTORS FOR THE U.S.A. AND CANADA:

Elsevier/North-Holland Inc.
52 Vanderbilt Avenue,
New York, N.Y. 10017

Library of Congress Cataloging in Publication Data

Nathanielsz, P.W.
  Fetal endocrinology.
  Bibliography: 36 pp.
  Includes index.
  1. Endocrinology, Experimental. 2. Fetus–
Physiology. 3. Sheep–Physiology. 4. Laboratory
animals. I. Title.
QP187.N28   599′.03′33   76-22194
ISBN 0-7204-0582-3

Printed in The Netherlands

Foreword

The last decade has seen a giant leap forward in fetal endocrinology to which concurrent developments in several areas have contributed. Advances in fetal surgery, particularly organ ablation and maintenance of chronic vascular catheters, and in radioimmunoassay of steroid and polypeptide hormones have been important, but perhaps the major single factor in this rapid progress has been the use of sheep rather than traditional small laboratory animals. The fetal sheep has yielded a veritable bonanza to the many investigators, including the author of this book, who have asked questions of it. The remarkable stability of pregnancy (much envied by those who work with primates), the relatively long pregnancy, the convenient size of the fetus and the docile mother, all combine to reward the experimenter.

The richness of the dividends from research in fetal sheep has proved to some investigators to be an irresistible temptation to move forward hastily with less than ideal animal preparations in order to uncover the next 'nugget'. Now the gold rush days are over and the treasures lie more deeply buried, to be discovered only as the result of patient, meticulous work. It is timely that Dr. Nathanielsz has written a book that forms a bridge from the past decades. He has used a thorough review of many aspects of fetal physiology as a background to a critical appraisal of experimental techniques and points the way to improved experimental design of relevance to a variety of species. The foundation to the experimental approach to fetal endocrinology was solidly laid by Professor Jost only a quarter of a century ago. Those of us following in his footsteps recognise our good fortune in having entered a young science in which the opportunity for the exercise of ingenuity is unrivalled and the examples of experimental design are so elegant.

G.C. Liggins

Postgraduate School of Obstetrics and Gynaecology,
The University of Auckland, New Zealand

Preface

There must be a good reason for writing a book. Rarely can it be said that none exist in the field. Certainly this is not true today in the field of fetal physiology. This may have been true in 1968 when Geoffrey Dawes synthesised the then current knowledge [162] but in 1976 we can look back on a decade of ever increasing activity with an escalating rate of publication of primary papers, review articles, symposia and textbooks. These publications have not ignored considerations of fetal endocrinology.

However, there are many pitfalls in the assessment of data in the area of fetal endocrinology which are unknown to those who have not worked in the field. In general, experimentation with the fetus imposes several major constraints. If the results are to have physiological significance, two experimental animals, mother and fetus must be maintained in as near normal condition as possible. This is particularly important when studying the endocrine function of the fetus since a major purpose of the endocrine glands is to respond to changes from the physiological norm.

Wherever possible the experimental work reported in this book has been selected on strict physiological criteria. Much of the material has come from experiments on the chronically catheterised fetal sheep. The advantages of this preparation are discussed in Chapter 2. Physiologists, endocrinologists and obstetricians cannot escape criteria of physiological normality using this particular preparation. A vast battery of investigations is available to enable scrupulous assessment of the stability of each individual fetal sheep preparation.

However, we must not lose sight of the fact that the sheep fetus differs in many fundamental aspects from other fetuses, including the human. For this, as well as other reasons, I have resisted the temptation to make this volume deal exclusively with the endocrinology of the fetal sheep and throughout have attempted to relate well-controlled data from the sheep to information gained in other experimental species and the human wherever conclusions can be made regarding general principles. Wherever information is available from other species and is not based on material of dubious physiological condition, if it throws useful light on the endocrine mechanisms the fetus employs during maturation, development and parturition, I have also endeavoured to include it. Certain areas have been neglected because the picture is too diffuse at the present time. There is little merit in the publication of an assorted set of experimental findings unless they present a cohesive whole which throws light on fundamental mechanisms. In addition, the endocrinology of carbohydrate metabolism and intermediary metabolism has not been considered in this book, since it is intended to deal with these areas in a subsequent volume.

The purpose of this book is to assist those who wish to become acquainted with both the recent advances and the problems of fetal endocrinology. It is my hope that it will be of use both to those working in the field and those obstetricians and reproductive physiologists who need to appraise the field critically but are not actively engaged in it.

Acknowledgements

This book is dedicated to all those whose real interests lie in the unravelling of the endocrinological mechanisms employed by the fetus in the later stages of gestation. I am deeply indebted to many people for the stimulus of their interest, discussion, collaboration and help over many years. To my colleagues and co-workers, Patricia Jack and Alan Thomas, without whose labours much of the work we have done together would have been impossible; to Fiona Bass, Carol Horn, Helen Shiers, Margaret Abel, Michael Parr and John Buckle; also to Marian Silver, who first helped me with fetal catheterisation. I am grateful to Elliot Krane for reading the original manuscript. I am also indebted to Geoffrey Dawes, Geoffrey Thorburn, Mont Liggins, Giacomo Meschia, Jean Wilson and Delbert Fisher, amongst many others for much stimulating discussion. I am particularly indebted to Mont Liggins for many hours of discussion and help with surgical and practical problems whilst working in his laboratory. I offer my sincere thanks to Dr. Alan Wallace, Dr. Geoffrey Thorburn, Dr. Guy Abraham, Dr. Ken Kirton and Dr. Colin Pierrepoint for various antisera. For her enthusiasm and commitment I would like to acknowledge Dr. Lesley Rees and also Sally Ratter for work we are doing on ACTH in the fetus.

My special thanks are due to Celia Perry for her untiring efforts with the manuscript and many other forms of assistance, without which this book would not have been possible.

Contents

Preface		v
Acknowledgements		vii
Contents		ix

Chapter 1   *General introduction* — 1

1.1	Physiological control mechanisms	1
1.2	Interrelationship of maternal, placental and fetal tissues	3
1.3	Fetal autonomy	5
1.4	Temporal sequence of the development of metabolic control in the fetus	5
1.5	Neurosecretion	6
1.6	Gradations of fetal hormone function	7
	1.6.1 The transition from fetus to neonate	7
1.7	Changes in tissue responsiveness during fetal life	9
	1.7.1 Gene expression in terms of protein synthesis	10
	1.7.2 Molecular basis of hormonal action at the cellular level	11

Chapter 2   *Methodology* — 13

2.1	Areas of investigation	13
	2.1.1 General anatomy	13
	2.1.2 Microscopical anatomy	14
	2.1.3 Biosynthesis of the hormone	15
	2.1.4 Hormone transport in blood	15
	2.1.5 Mechanisms of hormone action on target tissues	16
	2.1.6 Control of endocrine function	16

2.2	Techniques of investigation		17
	2.2.1 In vitro studies		17
	2.2.2 In vivo studies		19
2.3	When is a fetus not a fetus?		21
2.4	Criteria of assessment of sequential in vivo studies		22
	2.4.1 Pre-operative preparation		23
	2.4.2 Types of anaesthetic		23
	2.4.3 Post-operative period		23
2.5	The fetal primate model		28

Chapter 3 Endocrine function of the fetal testis 31

3.1	Introduction	31
3.2	Embryological development	31
3.3	Production of androgenic steroids in the fetus	32
3.4	Role of androgenic steroids in development	34
	3.4.1 Embryological structures which give rise to the gonads and reproductive tract	34
	3.4.2 Sex differentiation	35

Chapter 4 General features of the development and function of the fetal hypothalamo–hypophysial–portal system 45

4.1	Perfusion experiments	46
4.2	Functional studies following ablative surgery	47
	4.2.1 The rat	47
	4.2.2 The sheep	49
4.3	Setting of level of hypothalamo–hypophysial activity during fetal life	51

Chapter 5 The fetal thyroid. I. General features and experimental studies in ruminants 53

5.1	Developmental aspects	53
	5.1.1 Evolution of the thyroid gland	53
	5.1.2 Embryological maturation of the thyroid gland	54
5.2	The hypothalamo–hypophysial–thyroid–target tissue axis in the adult mammal	57
	5.2.1 Hypothalamus and adenohypophysis	57
	5.2.2 The hormones of the thyroid gland	59
	5.2.3 The role of T_3 and structure–activity relationships for iodothyronines	59
5.3	Experimental observations on fetal thyroid function in ruminants	61
	5.3.1 Sheep	61
	5.3.2 Calf	72

Chapter 6		The fetal thyroid. II. The thyroid in the rat, human and sub-human primate fetus	73
6.1		Function of the thyroid axis in the fetal and neonatal rat	73
	6.1.1	Development of thyroid function in the fetal rat	73
	6.1.2	Placental transport of iodothyronines in the rat and guinea pig	75
6.2		The human fetus and neonate	76
	6.2.1	Placental permeability to iodothyronines	79
	6.2.2	The role of iodothyronines in human cretinism	79
6.3		Experimental studies in the sub-human primate	81
6.4		The role of the fetal thyroid in development	83
	6.4.1	Growth	83
	6.4.2	Differentiation	83

Chapter 7		Structural and morphological development of the neurohypophysis	89
7.1		Chemical structure, synthesis and physiological properties	90
	7.1.1	Oxytocin	90
	7.1.2	Arginine vasopressin (AVP)	90
	7.1.3	Arginine vasotocin (AVT)	91
	7.1.4	Pathways of destruction of octapeptides	91
	7.1.5	Neurophysins	92
	7.1.6	Methods of measurement	92
7.2		Species to be considered	93
7.3		Stimuli which release neurohypophysial hormones	93
7.4		Oxytocin	94
	7.4.1	Placental transport of oxytocin	96
7.5		Arginine vasopressin	99
7.6		Physiological role of AVP in the fetus	100
	7.6.1	Release of AVP in response to hypoxia	100
	7.6.2	Fetal endocrine responses to haemorrhage	101
	7.6.3	Endocrine factors in the control of fetal body fluids	103
	7.6.4	Is AVP a corticotropin-releasing factor?	105

Chapter 8		Growth hormone, prolactin and placental lactogen	107
8.1		Hormonal similarities	107
8.2		Growth hormone in the fetus	108
	8.2.1	Sheep	108
	8.2.2	Human	114
	8.2.3	Some experimental observations in the sub-human primate	117
8.3		Prolactin	118
	8.3.1	Available measurements for hormone concentrations in the fetal pituitary and fetal body fluids	118

8.4	Placental lactogen		121
	8.4.1	Human studies	121
	8.4.2	Sheep studies	121
8.5	General conclusions		123

Chapter 9 Adrenocorticotropin 125

9.1	Adrenocorticotropin in the fetal sheep		125
	9.1.1	Significance of measurement of circulating plasma ACTH concentrations	125
	9.1.2	Fetal plasma ACTH concentrations in acute experiments	127
	9.1.3	Measurements of plasma ACTH concentrations in chronically catheterised fetal sheep	128
	9.1.4	Effect of adrenal growth on the increased production of cortisol by the fetal sheep	131
	9.1.5	Possible diurnal and other phasic changes in fetal ACTH secretion	133
	9.1.6	Calculation of fetal ACTH secretion rate	135
	9.1.7	Factors other than pituitary ACTH which may affect adrenal secretion of cortisol	136
	9.1.8	Possible positive feedback systems in the fetal hypothalamo–hypophysial–adrenal axis	138
	9.1.9	In vitro studies of changes in adrenal sensitivity to ACTH	141
	9.1.10	In vivo studies of adrenal stimulation by ACTH	141
	9.1.11	Mechanism of the increase in fetal adrenal sensitivity to ACTH	143
9.2	Function of the fetal hypothalamus		146
	9.2.1	Responses of the adrenal axis to various forms of stress	148

Chapter 10 The production and role of glucocorticoids from the fetal adrenal gland 151

10.1	Glucocorticoid biosynthesis in the adrenal gland		151
	10.1.1	Action of ACTH on the adult adrenal	152
10.2	Development of function in the fetal sheep adrenal cortex		153
	10.2.1	Growth of the fetal adrenal	153
	10.2.2	In vitro studies of fetal glucocorticoid secretion	153
	10.2.3	In vivo studies on plasma corticosteroid concentrations in late gestation	156
	10.2.4	Changes in fetal plasma corticosteroid concentrations at early stages of gestation	158
	10.2.5	Placental permeability to cortisol	159
	10.2.6	Changes in cortisol-binding proteins and cortisol BPR in fetal plasma during the last days of gestation	160
	10.2.7	Summary – four phases in the pattern of fetal plasma cortisol concentrations	161
10.3	Mechanism of action of glucocorticoids in the fetal sheep		163
10.4	The adrenal cortex in the human and the sub-human primate fetus		170
	10.4.1	Adrenal steroidogenesis	170

	10.4.2	Plasma cortisol concentrations in human amniotic fluid and human umbilical plasma	171
	10.4.3	Considerations of placental transfer of cortisol	173
	10.4.4	Pituitary control of the fetal adrenal	173

Chapter 11 *Fetal estrogens, progesterone and prostaglandins in the sheep and primate* 177

11.1	Sheep		177
	11.1.1	Estrogens in the fetal sheep	177
	11.1.2	Progesterone in the fetal sheep	182
	11.1.3	Prostaglandins in the fetal sheep	184
11.2	Human and sub-human primate		186
	11.2.1	Production of estrogens by the feto–placental unit in primates	186
	11.2.2	Relationship of fetal estrogens to changes in maternal plasma estrogen	187
	11.2.3	Progesterone in the fetal primate	188
	11.2.4	Prostaglandins in the fetal primate	189

Chapter 12 *Parturition and the feto–placental unit* 191

12.1	Introduction		191
12.2	The initiation, maintenance and successful completion of parturition in the sheep		193
	12.2.1	Historical background	193
	12.2.2	The central role of the fetal adrenal	195
	12.2.3	The input to increased fetal adrenal secretion of glucocorticoids	196
	12.2.4	The output side – how does cortisol bring about delivery of the fetal lamb?	200
12.3	The progress of pregnancy following fetectomy in experimental animals		208
12.4	The initiation, maintenance and successful completion of parturition in the human in the human and sub-human primate		210
	12.4.1	Historical background	210
	12.4.2	The input to the fetal adrenal in the primate	211
	12.4.3	The output side from the fetal adrenal	211
12.5	Future work		214

Appendix 217

Bibliography 221

Subject index 257

CHAPTER 1

General introduction

1.1 Physiological control mechanisms

The principles of fetal endocrinology are the same as those of endocrine mechanisms in the adult. This statement may seem self-evident. Both from a phylogenetic as well as an ontogenetic viewpoint, the endocrine system has developed as a control mechanism. The efficient functioning of the multicellular organism requires that the various activities of different tissue and organ systems should be correctly integrated. This co-ordination within the organism is regulated by two major control mechanisms, the nervous system and the endocrine glands. These systems may also be referred to as the neural and the humoral control mechanisms. Within each cell, different enzyme and subcellular units may be controlled by local factors such as the concentration of substrates or products of enzyme reactions or the availability of cofactors. However, when one cell wishes to signal an instruction to another cell at a distance, the message is either coded as nervous activity or in the shape of a chemical message, a hormone.

It is useful for us to have a working definition of a hormone. A hormone is a chemical compound secreted directly into the blood by an endocrine gland in response to a specific stimulus which may be nervous or blood-borne. The amount of hormone secreted varies with the strength of the stimulus. As a result of secretion into the bloodstream the hormone exerts its effect by modifying the activity of a target tissue(s) within the body.

Hormonally and neurally mediated control have been described, respectively, as chemically addressed and anatomically addressed control systems. The nervous system is termed an anatomically addressed control system since there are definite anatomical pathways – the nerve axons – down which the signals must pass. A particular tissue can only influence the activity of another tissue

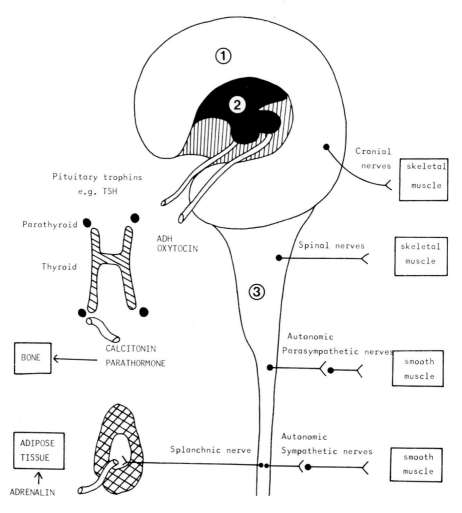

Fig. 1.1. Diagrammatic representation of the brain (1), hypothalamus and pituitary (2), spinal cord (3) and examples of neural and hormonal pathways.

if there is a direct nervous link between them. The nature of this link affects the coding of the message. For any specific fibre path the link is unidirectional, since synaptic connections between nerves demonstrate polarity (Fig. 1.1).

By contrast, in the endocrine system, the nature of the message is coded in the shape of the various hormone molecules and the ubiquitous distribution of the blood permits the access of the hormone to virtually all the tissues of the body. In certain regions there may, however, be important local variations in permeability across membranes. It is possible that certain hormones may not get into the cerebrospinal fluid in the fetus, newborn or adult. Similarly, the placental and fetal membranes (chorio-amnion and chorio-allantois) may influence the distribution of hormones between mother and fetus and even within the different body compartments of the developing fetus. It should be noted that the fetal membranes are highly vascular, even in those areas where they are not specialised to form the placenta. Consideration of transport of special molecules across non-placental chorion has not been investigated. These factors may place certain constraints on the principle of general access of hormones to all tissues.

1.2 Interrelationship of maternal, placental and fetal tissues

In our attempts to investigate the function of fetal endocrine systems we must make a necessary but arbitrary isolation of fetus from mother. Following fertilization, the zygote rapidly exhausts its own energy stores and is then dependent on the maternal tissues for nourishment, oxygen and the elimination of the waste products of metabolism. As its needs become greater, a specialised organ, the placenta, differentiates to provide the area of exchange between mother and fetus. The intricacy of this organ varies from species to species and its component tissues in the rat, sheep and human are shown in Fig. 1.2. These three species have been chosen since the bulk of carefully controlled experimental work has been performed on the rat and the sheep. The inclusion of the human requires no apology. Some of the techniques of experimental investigation of fetal endocrine function are designed to investigate the role of the placenta. Placental function can be studied using the in vitro systems and in vivo catheterisation techniques described in Chapter 2.

Fig. 1.2 shows that, in each species, part of the placenta is of fetal origin. These cells can be directly influenced by products of other fetal tissues since they are supplied with fetal blood. The fetal components of the placenta have the ability to secrete compounds which affect the activity of both maternal and fetal tissues.

Diczfalusy was one of the first to concentrate on the interrelationship of the fetal tissues and the fetal layers of the placenta in steroid metabolism (see Figs. 11.1 and 2). This interrelationship led to the concept of the feto–placental unit. According to this concept, steroids of fetal origin (for example, from the fetal adrenal) pass in the fetal blood to the placenta. In the placenta they are converted to other steroids. In certain species the placenta may be unable to synthesise some of these precursor steroids and may therefore be dependent on

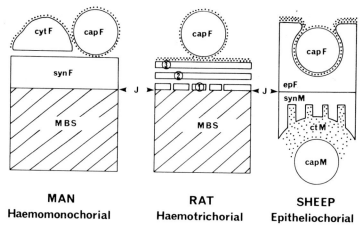

Fig. 1.2. Diagrammatic representation of the various fetal and maternal tissues which make up the placental attachment in the rat, sheep and human. capF, fetal capillary; capM, maternal capillary; ctM, maternal subsyncytial connective tissue; cytF, cytotrophoblast; epF, chorionic epithelium; J, fetal–maternal interface; MBS, maternal blood space; synF, syncytiotrophoblast; synM, uterine epithelial syncytium; ①, outer fenestrated layer of trophoblast, probably cellular; ②, ③, inner layers of trophoblast, probably syncytial. This figure was kindly produced by Dr. D.S. Steven.

the provision of the precursors from the fetus. In such instances, the fetal and placental function are interlinked and complement each other (see Chapters 10, 11, 12). Similarly the maternal tissues and the maternal placental tissues comprise a materno–placental unit, although the importance of the functional interrelationship is less clear in this instance.

Between mother and fetus, the placental barrier can display the following features:

1. It may be impermeable to the hormone in either the maternal to fetal direction (M → F), or the fetal to maternal direction (F → M), or in both directions. Impermeability in both directions appears to be the general rule in the sheep.

2. The placental tissues may modify the hormone molecule so that its action is changed. Thus, in the human, most of the cortisol which crosses from mother to fetus is initially converted to cortisone by the activity of an 11β-dehydrogenase enzyme. Since the fetus cannot reverse this step, this cortisone cannot be converted into an active form. Thus little active glucocorticoid appears to be transported from mother to fetus (see Chapter 10).

3. In theory, the placenta may allow the passage of a hormone from M → F or F → M. This process might be passive as down a concentration gradient, or it might be an active one.

1.3 Fetal autonomy

One of the most striking general principles to arise out of the vast amount of work in the field of fetal endocrinology over the last few years is that the fetus appears to be autonomous in regard to the control of its endocrine function. This observation should not come as too great a surprise. Since the fetus must control its own internal environment it must be able to maintain an independence from maternal hormonal changes. Changes in the maternal organism are, however, capable of altering fetal endocrine function. If the stimulus to the maternal endocrine system can be monitored by the fetus as a significant change in its own environment, then the fetus will respond. A good example is the effect of elevated plasma glucose concentrations in the maternal circulation. Such a change leads to secretion of insulin by the maternal pancreas. If the rise in maternal plasma glucose is great enough for sufficient glucose to cross the placenta and produce an adequate increase in fetal plasma glucose concentration, the fetal pancreatic β cell will be exposed to an increase in one of the major stimuli to insulin secretion. Dependent on several other factors, there may be a release of insulin from the gland and a rise in fetal plasma insulin concentrations. There does not however seem to be any significant passage of insulin across the placenta.

1.4 Temporal sequence of the development of metabolic control in the fetus

In the developing mammalian embryo and fetus, we may observe several different control systems:

(1) Cellular control systems   Initially, there are cellular control systems such as those which identify the specific functions of cells produced in the early divisions of the embryo [30].

(2) Inductors   Local factors known as embryonic inductors have effects on contiguous tissues. For example, the developing notochord produces substances which induce the development of the neural plate. Experiments involving the separation of notochord from presumptive neural plate by a porous membrane which is impermeable to these molecules, thus preventing them from reaching neural plate, result in the failure of differentiation of this structure. The actual biochemical nature of these inductors is not known but there are several examples at present under investigation.

(3) Simple feedback loops   Some hormone systems consist of a simple feedback loop such as that involved in the release of parathormone (Fig. 1.3). These systems generally begin to function later in fetal life than the two mentioned above.

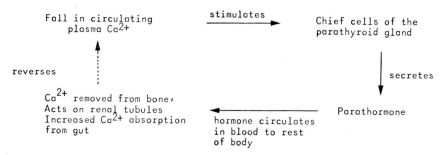

Fig. 1.3. A simple endocrine feedback loop. Decrease in plasma calcium concentration stimulates the chief cells of the parathyroid gland. These cells secrete parathormone. By several independent actions parathormone raises plasma calcium concentration again to the original level, thus removing the stimulus.

(4) Development of hierarchial control The simple feedback loop shown in Fig. 1.3 is ideally suited to the control of a basic feature of mammalian physiology such as the circulating plasma calcium concentration which is not exposed to rapid alteration or stimulation by either internal or external factors. Many such systems exist both in the adult and fetal mammal. Indeed the fetus is largely separated and protected by its mother's homeostatic mechanisms from any rapid changes in the external world.

However, the developing fetus must acquire the more complicated endocrine systems which will respond to external and internal stimuli at different levels. This development is especially concerned with the appearance of hypothalamic neuroendocrine systems. These systems are of two sorts and are shown in Fig. 1.4.

1.5 Neurosecretion

Nerve fibres must not be considered as passive channels whose only function is to conduct electrical impulses. They are also capable of secreting compounds from their terminals. The concept of synaptic chemical transmission is now firmly accepted. According to this concept nerve fibres release transmitter chemicals at synaptic junctions. These transmitters modify the state of polarisation of the post-synaptic membrane. It is now clear that similar processes occur in nerve fibres which do not end in a close relation to their effector organ. These fibres release transmitters – in this case hormones – which pass through small fenestrations in the adjacent capillaries and thereby enter the bloodstream.

A major location of such neurones is the various nuclei of the hypothalamus. We shall be particularly concerned with the development of the hypothalamic neurones which secrete the releasing factors, or releasing hormones, into the pituitary* portal system to control the activity of the pituitary. These neurones are part of a complex hierarchical control system which has several levels of activity (Fig. 1.4). In terms of preparation for an independent extrauterine existence we must consider the function of each individual level in these hierarchical mechanisms. Different features exist in the various systems such as the hypothalamo–hypophysial–thyroid and hypothalamo–hypophysial–adrenal axes.

1.6 Gradations of fetal hormone function

When considering fetal endocrine function we may conveniently consider the level of activity in each hormone system in terms of (i) basal function, and (ii) changes in the level of activity in the system at times of rapid alteration in the circumstances of the fetus such as maternal feeding, or its opposite, food deprivation, and such events as parturition.

1.6.1 The transition from fetus to neonate

This book is intended to cover certain closely defined areas. It cannot hope to be a comprehensive text of every molecule with endocrine function in the fetus. The terms of reference are:
(1) In general it is confined to the fetus.
(2) It will consider only molecules with unquestionable claims to hormonal status. In general there will be no consideration of locally active compounds or transmitters in the brain.
(3) It will be confined to a discussion of investigations performed predominantly in the sheep. The reasons for this are set out in Chapter 2.

However, eventually the fetus must be delivered. At this time it undergoes a remarkable transition within a short space of time to become a free-living,

---

* At various times, controversy has raged regarding the correct name for the vascular link between the hypothalamus and pituitary shown in Fig. 1.4. This system commences as a primary plexus of capillaries in the median eminence of the hypothalamus, larger vessels then run down the pituitary stalk. These vessels break up into a secondary plexus of capillaries within the substance of pituitary. It is clear that this vascular conformation is that of a portal system. The best known portal system in the body is the hepatic portal vein – which conducts blood from the gut to the liver. In the same fashion the portal system leading to the pituitary will be called the pituitary portal system in this book. An alternative name is the hypothalamo–hypophysial portal system. It should be noted that the hepatic portal system is not usually referred to as a gastrointestinal–hepatic portal system.

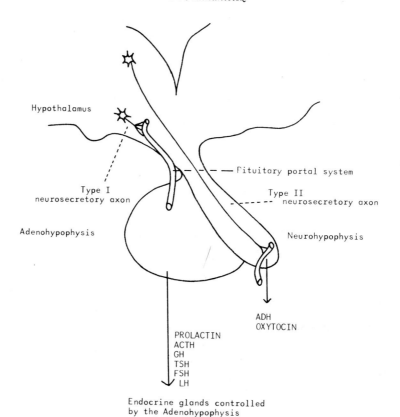

Fig. 1.4. Diagram of the essential neurosecretory features of the hypothalamus and the hormonal secretion of the adenohypophysis. There are two types of neurosecretion. Type I: hypothalamic neurones give off short axons that end on the primary plexus of the pituitary portal system. These fibres release hormones which control the adenohypophysis. Type II: longer axons pass to the neurohypophysis where they release arginine vasopressin (AVP) and oxytocin. The adenohypophysis itself releases hormones which control other endocrine glands.

relatively independent, neonate. It will be necessary for us to consider the endocrine systems which develop in preparation for this change. Fetal and neonatal physiology constitute a continuum. Fetal physiology is only meaningful when considered not only as a collection of physiological mechanisms capable of dealing with the special problems of fetal life but also as a preparation for an independent extrauterine existence leading up to activity as a mature adult, capable of reproducing the species.

1.7 Changes in tissue responsiveness during fetal life

The major advances in endocrine methodology which have enabled the measurement of hormone concentrations in peripheral plasma should not be seen as an end in themselves. It has previously been mentioned that the activity of an endocrine axis must be seen as a whole (Fig. 1.5).

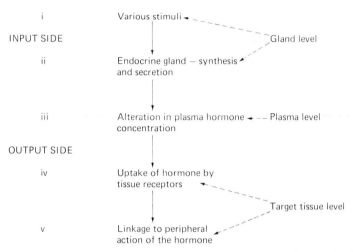

Fig. 1.5. Analysis of the various links between the stimulation of an endocrine gland and its action on peripheral tissues.

Measurements of increases and decreases in plasma hormone concentrations are now much easier to perform than they were a decade ago. As a result the tendency for endocrinologists to indulge in 'hormonal billiards' has been great. Whilst it daily becomes possible to measure the plasma concentrations of more hormones we must never lose sight of three major questions.

1. What are the physiological mechanisms which bring about an alteration in the circulating plasma concentration of the hormone?
2. What cellular changes are brought about by these hormone changes in enzymes, receptors, nucleic acids and other functionally significant cellular material?
3. How do these changes alter the physiology of the whole organism?

From the viewpoint of fetal development a clearer understanding of the actions of the hormone at the target tissue level (Fig. 1.5, steps iv and v), should be a major goal. As a result of great advances in knowledge of the basic mechanisms of hormone action, it would seem that changes in receptor concentration at the target tissue level may prove to be just as important as the changes in circulating hormone concentration. Changes such as these may be the basis of the alterations in gland sensitivity shown in the developing adrenal (Chapter 9). They may also explain why, under certain physiological conditions, alterations in plasma adrenocorticotropin (ACTH) concentration do not cause the expected increase in plasma cortisol concentration. Another important example from developmental biology is the observed change in sensitivity of rat cartilage to the growth-promoting effects of somatomedin (SM) at different periods of development. This observation may explain the apparent paradox that SM concentrations are low at times of rapid growth.

If we consider the role of hormones in the evolution of the many mammalian species there is the further possibility that the appearance of different chemical compounds in some tissues, glands or even plasma at various stages of fetal development may simply reflect phylogenetical factors rather than demonstrating that this actual molecule is of endocrine importance in the ontogeny of the particular fetus under investigation. The occurrence of the hybrid octapeptide arginine vasotocin in the fetal pituitaries of different species at certain stages of development (Chapter 7) and the observation that although the adult human intermediate lobe of the pituitary does not contain α-melanocyte-stimulating hormone (α-MSH) this molecule is present for a phase in the human fetal pituitary, are good examples of this consideration.

1.7.1 Gene expression in terms of protein synthesis

The replication of the genes takes place by a process of negative–positive image formation. In this system, pairs of purine and pyrimidine bases are specific for each other. The paired bases, thymine and adenine, guanine and cytosine, are formed by hydrogen bonds in the now familiar double helix structure. Splitting of the helix at mitosis is followed by production of two examples of the original chain by the two positive images formed as a result of the split.

Information transfer from the genetic code to produce a specific protein with a particular amino acid sequence requires two transcription phases. Initially, messenger RNA (mRNA) is produced, in which each individual amino acid is represented by a triplet nucleotide codon. This codon is further transcribed into a triplet anticodon on soluble or transfer RNA (tRNA). The tRNA sequence then undergoes translation into the amino acid sequence of the protein. These translation processes take place on the ribosome but other cytoplasmic factors are also of importance.

1.7.2 Molecular basis of hormonal action at the cellular level

Although great advances have been made in recent years in our knowledge of the biochemical mechanisms of hormonal action, there are still considerable gaps requiring elucidation. Several theories have been proposed in which hormones have been considered to act by the following mechanisms:

(1) Via intermediate, second messengers, particularly the cyclic nucleotides – cyclic AMP (adenosine 3′, 5′-cyclic monophosphate) and cyclic GMP (guanosine 3′, 5′-cyclic monophosphate).

(2) Changes in membrane permeability to important ions, particularly Ca^{2+}.

(3) By mechanisms which affect the expression of the target cell's genome. In this way hormones may modify the production of important cellular proteins, especially enzymes and carrier proteins.

The most extensively studied group of hormones in terms of their action at the cellular level are the steroid hormones. In all instances studied it appears that the steroid hormone binds initially to a receptor in the cytosol of the target tissue. This process is aided by the fact that they are small molecules of 300–400 daltons with a lipophilic structure. The receptor in the cell cytosol has a high affinity and specificity for the particular steroid which it binds. In some respects these cellular binding proteins are similar to the transport proteins which are responsible for hormone carriage in the blood. The steroid hormone may be transported to the nucleus of the cell and act without further modification. Alternatively, it appears in the case of testosterone, for example,

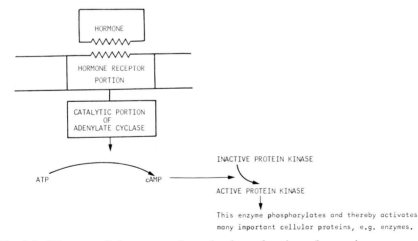

Fig. 1.6. Diagram of the proposed mechanism of action of some hormones on the plasma membrane of their target tissues to stimulate adenylate cyclase.

that the hormone molecule originally considered as the active principle is in fact converted into a more active metabolite (in this instance 5α-dihydrotestosterone) in certain tissues.

Polypeptide hormones are also bound to receptors on the cell surface. They appear to act by the first mechanism given above. It is considered by some authorities that the receptor for the hormone is part of a complex with the enzyme adenylate cyclase. The complex consists of the receptor subunit and the catalytic subunit. Attachment of the hormone to the receptor in some way activates the catalytic subunit (Fig. 1.6).

The various transduction steps mentioned above are the linkage of the hormonal message to the function of growing and differentiating cells.

CHAPTER 2

Methodology

2.1 Areas of investigation

The investigation of an endocrine system in the fetus employs exactly the same fundamental endocrinological methods as are used in the adult. The significant technical difference is the difficulty of access to the experimental animal, lying as it does surrounded by fluid-filled membranes within the maternal uterus, and related to the maternal circulation across the placenta. There are several general points which must be considered with reference to each endocrine gland in order to obtain a full picture of its development and activity. These are considered below.

2.1.1 General anatomy

The embryological origin of all the various tissues which together comprise a particular endocrine gland under study may have important bearings on the ability of the gland to function normally in fetal and adult life. Various methods of classifying endocrine glands exist. One method which has the advantage of pointing out certain similarities, refers to the tissue of origin of the endocrine cells – ectoderm, mesoderm or endoderm. Structural details of the surrounding tissues in the final site of the gland are also fundamental to functional endocrine interrelationships in several situations. The best example is the contact made by Rathke's pouch (the anlage of the adenohypophysis) with the floor of the hypothalamus during early embryonic life [198] (Fig. 2.1). This close apposition results in the possibility for local control of adenohypophysial function by hypothalamic hormones.

A further example of the results of location of origin and the migration of various component tissues on the function of endocrine glands is the influence

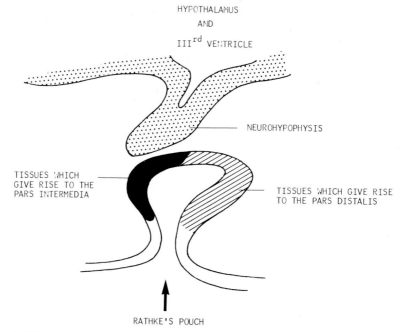

Fig. 2.1. Diagrammatic representation of the early development of the fetal pituitary. The adenohypophysis develops from an upgrowth from the roof of the mouth.

of steroids from the adrenal cortex on the induction of the N-methyltransferase enzyme system in adrenalin-secreting cells in the adrenal medulla. Since the venous drainage of the adrenal cortex is centripetal, the cells of the adrenal medulla are exposed to blood with a high glucocorticoid content. One of the many enzymes induced by glucocorticoids is the phenylethanolamine-N-methyltransferase in medullary chromaffin cells, responsible for the conversion of noradrenalin to adrenalin [461]. Details of embryological development will also assist in unravelling problems of innervation, arterial supply and venous drainage. Knowledge of the embryology of endocrine glands will also explain anatomical abnormalities due to aberrant development.

2.1.2 Microscopical anatomy

Extensive use of the light and electron microscopes has been made in the study of fetal endocrinology. Detailed study of both the endocrine cells and their target tissues can provide useful clues to the timing of the onset of endocrine function in a particular fetal endocrine gland. Microscopical investigation can also throw light on the general level of activity in the gland and is often useful for simultaneous study of the development of the target tissues of the hormone.

The use of histochemical, immunofluorescent and other sophisticated histological techniques are helping to solve problems such as the location of and the timing of development of specific enzymes, and the cellular localisation of different molecules.

2.1.3 Biosynthesis of the hormone

Biosynthetic pathways in endocrine systems have been extensively worked out in adult endocrine glands and in many instances this information is of use in the fetus. Thus, it can be seen in Chapter 5 that the successive steps in the synthesis of thyroxine (T_4) appear in the thyroid gland in an orderly sequence during early fetal life.

The synthesis of the iodothyronines requires adequate iodide and also a supply of the amino acid tyrosine. Failure to provide these precursors will prevent the appropriate synthesis of T_4. The production of steroid molecules based on the cyclopentanoperhydro-phenanthrene ring occurs in several tissues. The properties of steroids differ considerably according to the nature of the substituents on the basic ring structure (see Chapters 3, 10–12). A simplified diagram of the interrelationship of various adrenal, ovarian, placental and testicular steroids is shown in Fig. 3.1.

2.1.4 Hormone transport in blood

The small molecular weight hormones circulate in plasma in a reversible equilibrium involving various binding proteins (Fig. 2.2). A portion of the circulating hormone molecules are bound to various plasma proteins; the remaining molecules are free in the plasma. The proportion of the hormone in the free state varies between different hormones. In addition, the proportion for any one hormone may change under different physiological conditions. It is

```
THYROXINE  +  THYROXINE    ⇌    THYROXINE ∿ THYROXINE
              BINDING                       BINDING
              GLOBULIN                      GLOBULIN

- Free, or    (TBG)                - Bound T₄
  unbound
  T₄
```

Fig. 2.2. Binding and dissociation of thyroxine with its major carrier protein in plasma, thyroxine-binding globulin (TBG).

generally considered that the concentration of free hormone in plasma is the physiologically important fraction of the circulating hormone, being directly available to the target tissues.

The protein-bound hormone may constitute a reservoir of hormone in the blood. A further possible function in the adult may be to lessen the proportion of the hormone which is filtered at the renal glomerulus. Both T_4 and cortisol are bound to several different plasma proteins. It may be that some of the hormone–protein complexes constitute a mechanism whereby the hormone crosses plasma membranes or epithelia and gains access to specific sites.

From the practical point of view it is important to consider the possible influences in the changes in hormone binding which occur in the fetal period. Two particular instances are discussed below. In Chapter 5, attention is given to the changes in T_4 binding which may explain various discrepancies in measurements of total plasma T_4 in the immediate prenatal period. Alterations in total plasma T_4 may take place without any change in the total concentration of free hormone if the percentage of the hormone which is non-protein-bound changes as well [593]. Secondly, it is necessary to investigate the exact extent and the importance of the 10-fold increase in fetal plasma cortisol concentration which occurs before birth. It must be shown that the increase in fetal plasma cortisol is indeed accompanied by increased concentrations of free hormone and an increased turnover of the hormone (Chapter 10).

2.1.5 Mechanisms of hormone action on target tissues

The various possible mechanisms of action of hormones at the tissue level are discussed in the preceding chapter. With each endocrine system investigated we need to look at the metabolic actions controlled by the hormone.

2.1.6 Control of endocrine function

In the case of each individual endocrine gland investigated in the fetus it is necessary to recall the considerations discussed in Chapter 1 in which the control function of endocrine glands was defined. Two extremes of complexity were discussed – the chief cell of the parathyroid gland which responds directly to blood calcium concentration and the hierarchical system such as the thyroid axis (Figs 1.3 and 5.1).

For adequate function of each endocrine control system in the fetus all the links in the system must have differentiated. The intracellular synthetic apparatus of the endocrine cell and that of its target tissue must have differentiated completely. In addition, the hormone receptors must be present on the target tissue and on the endocrine cell itself. These problems are considered in detail in relation to adrenocorticotropic hormone (ACTH) and adrenal sensitivity in Chapters 9 and 10.

Experimental investigation of complex hierarchical endocrine control systems such as the thyroid or adrenal axes is in essence the investigation of the function of several endocrine glands. Experimental observations are often capable of multiple interpretation. In order to obtain a clear picture, several hormones, both steroid and polypeptide, must be measured simultaneously and experimental design must be carefully prepared. The elimination of possible influences at different levels of the system can take years of multidisciplinary investigation. The investigation of the activity of the fetal hypothalamo–hypophysial–adrenocortical axis illustrates these problems excellently.

2.2 Techniques of investigation

2.2.1 In vitro studies

Removal of tissue from the mammalian organism for in vitro study has certain very real advantages. This is especially true for studies of fetal function. When placed within a convenient container on the laboratory bench, fetal tissue is much more accessible than it is in situ. The addition of radioactively labelled precursors, inhibitors and other reagents, some of which may be toxic when administered to the whole animal, can be effected easily by the investigator. However, the very ease of addition masks the fact that these experiments can only suggest what may happen in vivo. They can never state what actually does happen in vivo. The in vitro system lacks the ability to reproduce the availability of precursors and the effects of physiological changes in blood flow. It cannot mimic the results of alterations in vascular permeability and changes in blood gas tensions. Many workers completely ignore the fact that, in vivo, fetal tissues are exposed to very different blood gas tensions from the adult. In addition, as a result of the peculiarities of the fetal circulation (oxygenation in the placenta as opposed to the lung, and the presence of various vascular shunts), gas tensions and pH vary considerably in different regions of the fetal circulation (see Table 2.1).

Table 2.2 shows values obtained from fetal sheep for several fetal hormone concentrations in peripheral plasma around the time of birth. In many instances in vitro studies have been conducted with hormone concentrations several orders of magnitude greater than these. While we must bear the pitfalls of this form of investigation in mind, in vitro work can yield much useful data in spite of the limitations. Perfusion of fetal newborn lamb and adult sheep adrenal

Table 2.1.
Blood gas tensions and pH of whole blood in different regions of the circulation of the fetal lamb under different physiological conditions.

	Reference	Gestational age (days)	pH	pO_2	pCO_2
Umbilical artery	[137]	121–140	7.35	22.8	45.7
Umbilical vein	[137]	121–140	7.40	34.8	41.5
Carotid artery	[165]	94–141	7.32	22.0	47.5
Inferior vena cava	[138]	9 days before parturition	7.39	25.4	49.0
Inferior vena cava	[138]	1 day before parturition	7.38	25.0	49.0

slices in a system with 300 ng/ml of synthetic ACTH (Synacthen) solution being continuously pumped through the system produced the interesting observation shown in Fig. 2.3, taken from the work of Madill and Bassett [403]. It is clear that the sensitivity of the adrenal to a fixed concentration of ACTH increases about 10-fold over the last ten days of intrauterine life. This observation is of great significance and will be discussed again in relation to the function of the fetal adrenal cortex. In vitro demonstration of this very important point did however require superfusion with ACTH concentrations 1000-times those

Table 2.2.
A comparison of absolute and molar concentrations of various hormones in fetal lamb plasma on the last day of gestation. Glucose concentrations in the fetal lamb and the adult human are given for comparative reasons. For ACTH, 1 pg: 1.3×10^8 molecules. The sensitivity of the cytochemical assay is 5 fgrams or 6.5×10^5 molecules.

	Molecular weight	Plasma concentration (ml^{-1})	Molarity
Cortisol	362	150 ng	410 nM
Aldosterone	360	250 pg	690 pM
Thyroxine	777	50 ng	64 nM
Triiodothyronine	651	500 pg	770 pM
Thyrotropin	28,000	1 ng	35 pM
Adrenocorticotropin	4600	1 ng	217 pM
Oxytocin	1000	10 μU or 5 ng	5 nM
Antidiuretic hormone	1000	10 μU or 5 ng	5 nM
Blood glucose	180	150 μg	830 μM
	Adult human	800 μg	4.4 mM

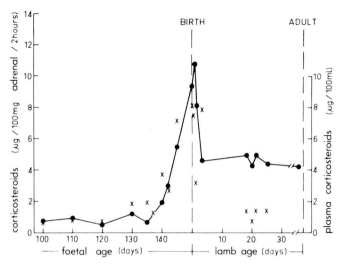

Fig. 2.3. Adrenal response to ACTH in vitro. The mean rate of corticosteroid release (solid symbols) by perfused adrenal tissue slices from fetal and postnatal lambs, during the 2nd and 3rd hours of perfusion when Synacthen (0.3 µg/ml) was added to the medium. The plasma corticosteroid concentrations (×) of the lambs from which adrenal tissue was obtained are also shown. Each point is the mean of results for 2 or more lambs. Reproduced with permission from [47].

circulating in the fetal lamb. Even in the presence of these high ACTH concentrations, the maximal glucocorticoid output was only equivalent to 1.2 mg/g/day. This is equivalent to about a twentieth of the secretion rate expressed per gram of adrenal per day in vivo, showing that tissues generally lose some of their responsiveness when studied in vitro.

2.2.2 In vivo studies

Animal material used for the investigation of fetal endocrinology may conveniently be divided into two groups:
1. Material taken under anaesthetic or at time of death for single point analysis.
2. Serial samples taken from the same animal over a period of time.

Single-point analysis of plasma composition
The majority of data concerning hormone and other related measurements in fetal mammals has been obtained from fetuses removed from the uterus after death of the mother, or whilst the mother is anaesthetised or after abortion of

the fetus. In certain small species these methods are the only practical ones. In the human, material can only be obtained after abortion, spontaneous or induced; caesarean section under anaesthesia, or after vaginal delivery. As discussed below, in none of these situations can the material be considered as exactly representative of fetal material. In experiments with pregnant sheep it is probably preferable to obtain material under epidural or other regional anaesthesia than whilst the mother is under a general anaesthetic.

Sequential in vivo studies
From the outset, the reader should exercise an element of caution with regard to the interpretation of this valuable data. In species in which it is possible to compare single-point results obtained under anaesthesia with sequential samples obtained from stable, chronic, unanaesthetised preparations, several fundamental differences have been observed. This is especially true of endocrine mechanisms. Practical and ethical considerations preclude a direct comparison of observations from single-point and chronic experiments in most species but the lessons learned from instances in which such comparisons can be made should always be borne in mind. In some situations no difference has been observed between chronic and acute experiments (see below) but the limitations of single-point analysis must be constantly considered. This caution must be exercised particularly with respect to endocrine mechanisms which respond rapidly to metabolic and blood gas changes. A further reference will be made to single point analysis after consideration of sequential sampling regimes in the chronically catheterised fetus.

Single-point analysis of tissue composition
Measurement of hormone and metabolite concentrations in plasma may not reflect changes occurring in particular tissues. Certain hormones may be formed within tissues from precursors (see Chapter 3). Several systems in which care must be taken in the interpretation of plasma measurements are discussed at length in the book. In view of the fact that the 'active' hormone may be formed within the target tissue it is very necessary to ask, what is the significance of circulating prostaglandin, progesterone and triiodothyronine concentrations in comparison with the concentrations within the tissues, if we knew them? It is becoming clear that more knowledge is required of the various intermediates in the synthesis of estrogens but it is highly likely that some of these compounds are never secreted into the blood. Measurements of plasma hormone concentrations may miss physiologically significant changes in such systems. Investigations of systems such as these, and the enzyme steps involved, will require excision of tissue from animals in various physiological states with the

subsequent investigation of the activity of different hormones and enzymes within these tissues.

Measurement of gland content

Endocrine glands differ considerably in the amount of stored hormone they contain within the gland cells. Steroid-producing cells, particularly those of the adrenal cortex, contain very little stored hormone. In contrast, the various cell types of the adenohypophysis do contain stored secretion. Even within the adenohypophysis there is considerable variation in the different cell types. The somatotrophs which synthesise growth hormone appear to have the greatest capacity to store hormone, at least in the fetal sheep.

The hormonal content of any endocrine gland represents the final balance of several processes which can be expressed mathematically in the following equation:

Content of hormone in gland = Hormone synthesised − hormone secreted − hormone degraded within the cell.

Attention to this formula will avoid the incorrect use of the word secretion with reference to endocrine glands. Synthesis refers to the production of the hormone whereas secretion refers to the release of the hormone from the gland. Too often the word secretion is used to refer to both these processes.

This equation demonstrates the inability to assess the level of function of an endocrine system by simply measuring the gland content of the hormone which may fall as the result of decreased synthesis or increased secretion occurring independently. Usually the stimulus to secretion is also a stimulus to synthesis. However, in this book, gland contents of hormones during fetal life will be used simply to define that the gland has the capacity to produce the hormone at that stage of development. No inferences will be made as to the level of activity from such data.

2.3 When is a fetus not a fetus?

It is necessary to consider a widely exhibited misuse of the word fetal. In many publications it is customary to refer to cord blood concentrations of hormones and metabolites as fetal concentrations. The argument as to when a fetus ceases to be a fetus may sound academic and semantic. This problem could be resolved with ease if, under all conditions, the exact condition of the experimental material is supplied. Details such as gestational age, timing from surgery and type of anaesthesia are essential to the interpretation of any experimental results. The various influences which have acted on the fetus can then be taken into consideration. This is more important than the actual designation given to the values as being those for fetal or neonatal plasma. Whether or not the terminal changes during the delivery of the fetus down the birth canal occur in an animal which is dependent on the mother for oxygen and other nutrients and can thus be considered a fetus or whether it is now a free-living organism, will depend on several factors. A factor of particular importance which will greatly affect the function of the fetus is whether the maternal uterine and placental blood flow is adequate or is being compromised by uterine contraction. Placental blood flow will be a limiting factor even if the umbilical cord and placenta are still attached to the uterus. Endocrine features of the fetus under these conditions are not representative of the stable situation which existed in utero several days prior to the onset of delivery. Similarly, after delivery, even before the cord is sectioned, the postdelivery condition is determined by the new environment in which the animal finds itself. The onset of respiration in a normal unassisted vertex presentation, in both the human and the lamb, often occurs before the trunk is fully delivered. If it does not start then, respiration commences when the cord is broken. Such pronounced changes mean that the animal should now be considered as a neonate. If we designate it as such, the use of the word fetal should not mislead us into thinking that this new situation has much similarity to that which existed in utero.

2.4 Criteria of assessment for sequential in vivo studies with special reference to the chronically catheterised fetal sheep

In his fundamental studies on the actions of vitamin A, Sir Gowland Hopkins exposed the same rats to both deficient and adequate diets. The diets were switched during the experiment, thus exposing each animal to both the experimental protocols and enabling the use of the individual animal as its own control. In adult mammals, endocrine investigations involving changes in plasma concentrations are usually preceded by a period of observation of the variables under examination in the resting state. It is also possible to revert to a control period after the test. For similar studies on fetal endocrinology, the sheep fetus is the best experimental model available at the present time, for the reasons listed in Table 2.4.

The species which have been used for sequential studies of fetal endocrine function are the sheep, goat and horse [139, 439]. Even within this restricted group, various species differences occur. For example, of these species only the goat has a monogastric digestive system. Care must, therefore, be taken with relation to development and function of the digestive system and gastrointestinal

Table 2.3.
Data of some currently available methods of assessing normality in the fetal sheep preparation.

Fetal blood gas tensions:
  These vary according to the site of measurement, see Table 2.1. Acceptable values must be determined in each centre of investigation [140]. It is to be hoped that indwelling electrodes will soon be available for continuous monitoring.

Fetal pH

Fetal plasma glucose concentrations:
  These are very dependent on the level of maternal plasma glucose concentration. Results differ according to the different methodologies used. The most specific technique is that using glucose oxidase. For a review of this problem, see Shelley [516].

Fetal respiratory movements:
  The frequency and type of fetal respiratory movement is affected by level of fetal oxygenation and can be correlated with various other parameters of fetal wellbeing.

Fetal heart rate and blood pressure:
  Changes in the fetal cardiovascular system which are reflected in fetal heart rate and the normal beat to beat variation in the cardiac cycle are used routinely in fetal monitoring. From the point of view of the endocrinologist it may prove that these are not the earliest manifestations of fetal compromise.

---

endocrinology when ruminant fetal models are used. Various important differences also occur in placental structure in these species. It has been elegantly shown by Silver, Comline and Steven [525] that blood gas transport across the placenta shows characteristic differences between the sheep and the horse. Placental endocrinology also differs. Thus, the sheep placenta is the major source of progesterone in the later part of pregnancy, whilst in the goat pregnancy does not continue after the removal of the corpus luteum [553].

As mentioned in the Preface, considerable emphasis will be placed in this book on sequential data obtained from the chronically catheterised fetal sheep preparation. Even with the fetal sheep, criteria for normality of the preparation are disputed by different workers. Several different variables have been monitored in order to assess fetal condition. Table 2.3 contains a list of some of the forms of data currently available for the assessment of normality in the fetal sheep preparation. The discerning reader will reserve judgement of basic papers which do not refer to control catheterisation details, and that do not contain basic data for the condition of the fetus. Normality of blood gases and other fetal plasma features do not prove normality of the fetus, but they do to some extent exclude the possibility that the fetus is severely stressed.

The different catheterisation techniques used all have certain advantages and disadvantages. However, very little attention has been paid to the consideration as to when a preparation is stable enough post-operatively to commence experiments. These questions have been discussed before and attempts at certain guidelines made [439, 516]. It is possible to maintain catheters for up to 53 days with the subsequent delivery of a live lamb [412].

2.4.1 Pre-operative preparation

Since fetal plasma glucose concentrations reflect maternal plasma glucose concentration, pre-operative maternal starvation may produce a fall in fetal plasma glucose. When general anaesthesia is used, it is common practice to deprive the ewe of food for 48 hours [414] or 24 hours [137]. With epidural anaesthesia some workers have used overnight fasting [516] or up to 48 hours of feed restriction [419]. It is, however, probably unnecessary to withdraw food at all when using epidural anaesthesia and this should prevent the tendency of maternal and fetal glucose concentrations to fall.

Progesterone treatment before surgery has been used by some workers [137, 419, 414], in order to prevent uterine contraction. In the light of the findings of Liggins and co-workers [391], the dosage of progesterone used would probably have been inefficacious in preventing labour.

2.4.2 Types of anaesthesia

The method of maternal anaesthesia employed not only affects the overall success rate for the surgery but will be an important factor in the length of time it will take for a preparation to stabilise post-operatively. It is very probable that general anaesthesia has a greater depressant effect on placental circulation than epidural anaesthesia. In addition, general anaesthesia will depress maternal respiration, thus affecting fetal oxygenation. No direct comparison exists between general anaesthesia and epidural regional anaesthesia but my experience with both systems leads me to the unequivocal statement that epidural anaesthesia produces better results.

2.4.3 Post-operative period

The period of time required for an individual fetal preparation to stabilise will depend on the pre-operative preparation and the type of anaesthesia. Considerable data is available on the changes in various indices of fetal function post-operatively but there are no direct comparisons between different surgical regimes. Various features are discussed below.

Feeding and carbohydrate metabolism

Shelley [516] points out that voluntary food restriction is a feature of the response of the sheep to strange conditions. In a study of 13 fetuses in 10 ewes prepared for surgery by an overnight fast and catheterised under epidural anaesthesia, maternal food intake did not return to normal for 4–5 days. Maternal and fetal blood gas values had stabilised within 24 hours but it took 3–5 days before fetal plasma glucose, fructose and lactate concentrations had reached stable levels. Such observations as these have an important bearing on studies of fetal energy metabolism [524].

Fetal endocrine changes

The stress of surgery results in an elevation of fetal plasma ACTH concentrations which return to normal in 24–48 hours (Table 9.1). The unresponsiveness of the fetal adrenal cortex to ACTH at the time that catheterisation is usually performed (90–130 days) is a protection against a marked increase in fetal plasma cortisol as a result of this ACTH secretion. The release of ACTH by the sheep fetus in response to stress is discussed in Chapter 9. Release of fetal AVP may be the explanation of the reduction in the rate of production of urine by the fetus post-operatively [251]. Fetal mineralocorticoid action on the chorio-allantois has also been considered to be the cause of the changes in allantoic fluid composition which follow starvation and surgery [414]. Mellor and Slater [414] make the very important observation that during the recovery period after fetal catheterisation, maternal and fetal plasma ionic composition returns to normal within three days but stability of fetal fluid composition is not achieved until after seven days. It is therefore clear that normality expressed by measurement at one site does not mean that normality has been achieved at all locations in the fetus.

Fetal respiratory movements

The work of Dawes and his colleagues has demonstrated that irregular rhythmic movements of the fetal chest wall occur in relation to rapid low-voltage EEG activity recorded from biparietal electrodes [163, 165]. Maternal general anaesthesia inhibits fetal respiratory movements and converts rapid low voltage EEG activity to slow high-amplitude activity. In addition the proportion of the day that the fetus spends in rhythmic respiration is greatly depressed post-operatively. When two days has elapsed after surgery a stable pattern has been established in which about 30% of the day is spent in rhythmic breathing [165].

Blood gases and pH measurements

The commonest variables used for monitoring fetal well-being are PO_2,

PCO_2 and pH measurements. Table 2.1 shows the data which have been accumulated for fetal carotid blood, inferior vena caval blood, umbilical venous and umbilical arterial blood by various workers. As mentioned above, Shelley has demonstrated that normal blood gas tensions can be observed within 24 hours of surgery whereas concentrations of several plasma metabolites have not stabilised at this time. In addition is should be recalled that Mellor and Slater [414] have shown that for several ions, plasma concentrations observed post-operatively may not reflect the situation at other sites in the body. It may also be incorrect to believe that normal fetal blood gases and pH invariably indicate a normal preparation.

In one study, exteriorisation of sheep fetuses whilst the mother was given low spinal anaesthesia resulted in an increase in umbilical venous PO_2, but this was achieved in the presence of a reduction in umbilical blood flow. The changes in the fetal circulation were not reflected in the values obtained for fetal blood gases and pH [280]. Redistribution of blood flow has been shown in response to hypoxen ia in the actue and chronic fetal sheep preparation [131, 164]. These vascular changes may be homeostatic mechanisms directed to maintaining cerebral and myocardial blood flow. From the viewpoint of the experimenter, wishing to interpret changes observed in his particular fetal sheep preparation, these observations should introduce an element of caution. Rather than being an index of stability, apparent normality of peripheral values in any homeostatic system may reflect successful compensatory responses to adverse conditions at other sites, particularly in the central nervous system.

In conclusion, it is to be hoped that reports of work with the fetal sheep will contain at least the following details, or references to where they may be obtained:

1. Preoperative preparation of the animals.
2. Gestational age at operation*.
3. Details of surgery and drugs, antibiotics, etc.** administered.
4. The report by Comline and Silver [137, 138] of the details of the chronic sheep preparation in their own hands is an object lesson in the provision of baselinc data. Such baseline data should be available for comparison with results obtained from different types of investigation.
5. Time interval between surgery and the commencement of sampling.
6. Some indices of fetal function such as pH, blood gases, plasma glucose

---

* Gestational age can be confirmed by limb X-ray [382].
** In 3 pregnant ewes treated with either penicillin G and dihydrostreptomycin or ampicillin, changes were observed in amniotic or allantoic fluid composition [417]. In 2 of the 3 instances the changes were not reversible.

concentration, etc., so that both their absolute values and any trends during the experiment, may be assessed.
7. In many experiments related to parturition, uterine contraction records are necessary to assess the level of uterine activity.
8. The ultimate fate of the lamb should be given. It is not unheard of for the statement that 'the lamb was delivered by caesarian section' to refer to the fact that this was a dead lamb although that information is not given.

The author would be the first to acknowledge that these are standards that he has not always reached himself. However, it is to be hoped that information such as that given above will become more routine in all publications referring to the fetal sheep. Finally, it is clear that in the last resort, criteria of acceptability will be governed by criteria of possibility. Table 2.4 lists some of the advantages and disadvantages of the fetal sheep model.

Table 2.4.
Advantages and disadvantages of the fetal sheep model.

---

1. Economic considerations. The use of the sheep in large numbers for meat and wool production means that the cost is much less than for pregnant females of more exotic species.
2. Availability of suitable pregnant animals. There are various well tried hormonal methods of inducing estrus in the sheep [494, 499]. This ensures that pregnant sheep can be obtained for most of the year in both the northern and southern hemispheres.
3. Past obstetric history. It is possible to obtain ewes whose past obstetric history is known.
4. Size. Depending upon the breed, the fetus weighs 2.0–6.0 kg at term. This has several advantages.
    (a) *Blood sampling*. The relatively large blood volume permits adequate plasma to be removed for sequential sampling from the same animal in most investigations. For a discussion of the effects of repeated blood sampling see Mott [427]. The availability of sensitive saturation analyses and micro-chemical techniques mean that it is possible to follow plasma concentrations of several hormones and metabolites simultaneously.
    (b) *Ablative surgical techniques*. Removal of various endocrine glands and section of important nervous structures can be performed.
    (c) *Recording of electrical activity*. Recording electrodes can be placed at various locations to record electrocardiogram (EKG), electroencephalogram (EEG) [77] and electromyogram (EMG) activity.
    (d) *Collection of body fluids*. Catheters can be inserted into various body cavities for collection of body fluids such as urine, amniotic and allantoic fluid. Flowmeters can also be implanted to measure flow rates.

5. Twin model. In a twin pregnancy, both fetuses can be catheterised. Surgery or the intravascular administration of various agents can be undertaken on one fetus and the other used as a control. Occasionally there are small areas of vascular connections between the two placentae in twin pregnancy. If such connections exist, the preparation cannot be used for isotope studies or hormone infusion. Injection of radioactively labelled polypeptides (such as growth hormone) which are known not to cross the placenta can be used to test for any possible vascular confluence.
6. Disadvantages. It would be surprising if there were not some disadvantages:
    (a) There are some important endocrine species differences between the sheep and other mammals.
    (b) There are well documented species differences in intermediary metabolism.
    (c) This preparation is not very suitable for single point analysis of tissue concentrations of hormones, particularly from the point of cost.

2.5 The fetal primate model

This book contains much less data from primate preparations than from the chronically catheterised fetal sheep. The problems of animal management, surgery and sampling in the fetal primate model are discussed in several papers [122, 123, 436]. A considerable amount of data is available regarding the changes in the maternal endocrinology of the primate during pregnancy [1, 53, 84, 109, 110, 116, 261, 262, 364, 365, 462] but little information other than single point analysis or observations under anaesthesia is currently available with reference to the endocrinology of the fetus. The purpose of this book is to discuss the information available from long-term, detailed studies, to a large extent carried out on individual animals in which monitoring of fetal well-being, uterine activity and a wide range of maternal factors can be carried out simultaneously. This is not yet possible with the primate.

In the rhesus monkey *(Macaca mulatta)* the placenta is generally composed of a large (or primary) disc and a smaller (or secondary) disc. Insertion of the umbilical cord is into the primary placenta and interplacental vessels run directly from the cord to the secondary disc between the amnion and chorion. It is therefore possible to incise the uterine layers and the chorion to expose these vessels without entering the amniotic cavity. A T-tube catheter can be placed in the fetal artery or vein to permit sampling of fetal blood through the side-arm without having to stop the flow through the vessel. However, whilst these catheters can be used for sampling fetal blood under anaesthesia, it has not been possible to produce a stable chronic preparation with fetal vascular catheters. The maternal monkey is very much more agile than the ewe and greatly resents any form of restraint.

Wherever possible and informative, fetal primate data is included in this book. As a result of differences in endocrine physiology, different animal models will have varying suitability for extrapolation to the human. Even amongst the sub-human primates there are considerable differences [506]. Such extrapolation is not the fundamental purpose of this book. The author hopes that useful information and some direct clues will come from the animal models which we can currently control within well-defined limits. Hopefully some of the major problems of chronic primate preparations will soon be solved and more hard data will be available from fetal primate material.

CHAPTER 3

The fetal testis

3.1 Introduction

It is clear from several lines of observation that the testis has a more important and active role to play in fetal development in the male fetus than has the ovary in the female fetus. Fetal castration in either sex results in the development of female features. From this observation we may derive the primary rule of sexual differentiation, that development which occurs in the absence of the male influence (whatever the actual agent may be) is female in character. For female development to occur there is no requirement for a positive factor derived from the ovary. Put in another way, the various male factors convert the female form of development into a male form. However, as discussed in Chapter 11, considerable concentrations of estrogens can be demonstrated in the fetal circulation at different stages of fetal life. The source of these estrogens is the placenta and it is possible that they may play a positive role in the differentiation of the female.

For these reasons we shall spend much more time considering fetal testicular function than ovarian function. The most carefully controlled experimental data has come from the mouse, rat, rabbit, sheep, calf, guinea pig and human fetus, and we shall restrict our considerations to these species.

3.2 Embryological development

The testes develop from the urogenital ridge which is a mesenchymal thickening on the dorsal wall of the coelom. These cells eventually give rise to both renal and genital structures. At an early stage the primordial germ cells migrate into the gonad from the yolk-sac endoderm. Initially the developing gonad is indistinguishable in the male and female. Two features subsequently differentiate the testis from the developing ovary. Well-developed, branching strands of cells, the testis cords, appear together with the initial appearance of the surrounding fibrous capsule of the gland – the tunica albuginea. As described below, the duct system of the testis is the remnants of the mesonephric duct which have only a transitory relation to the developing kidneys.

3.3 Production of androgenic steroids in the fetus

C19 steroids with known androgenic function are produced in the testis of the male fetus, in the adrenal glands of male and female fetuses and as intermediates in placental biosynthesis. Fig. 3.1 shows the overall relationships of the hormones to be considered in this chapter to other steroid hormones. Fig. 3.2 shows the major enzyme pathways we are concerned with in relation to the synthesis of androgens.

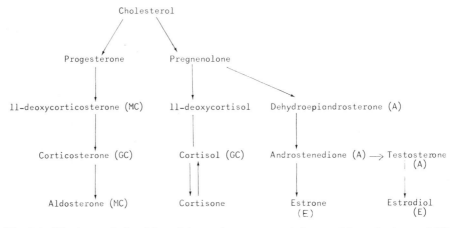

Fig. 3.1. The interrelationships of the major compounds in steroidogenic tissues. MC, steroids with mineralocorticoid activity; GC, glucocorticoid activity; A, androgens; E, estrogens.

Various approaches have been made to demonstrate the activity of the fetal testis at different stages of gestation. Specific histochemical studies to follow the appearance of activity of important enzymes in steroid biosynthesis have been made. From these and in vitro experiments on the conversion of major precursors, especially Δ^5-pregnenolone, and measurement of fetal blood samples taken at hysterotomy, it is clear that the testis is capable of producing various androgens in the fetal rat [585], rabbit [588], sheep [29], calf [425] and human

[3]. A common feature of several species in which the required data is available is that whilst there is no difference between the circulating plasma concentrations of androgens at term in male and female fetuses, measurement of different indices of androgen production at a much earlier phase at which sexual differentiation is occurring (see below) show that there is a period of increased testicular synthetic activity. Where measurable, circulating androgen concentrations are higher in the male than the female at this time [3, 425].

It can therefore be demonstrated experimentally that fetal androgens are synthesised and present at the critical time to undertake the roles described below in sex differentiation. Further than this we cannot go at the present time. These early stages of development are not amenable to the type of experiment using vascular catheterisation that have demonstrated that fetal androstenedione and testosterone are present in fetal sheep blood at the time of delivery and that these levels increase before delivery [388, 539]. Similarly, the role of

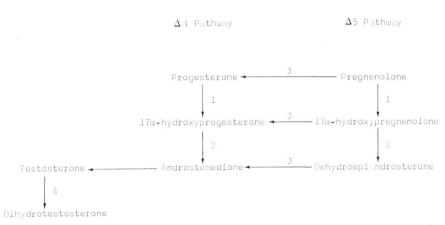

Fig. 3.2. Conversion steps by which the various androgens are related to each other and their major precursors. The major enzymes are (1) 17α-hydroxylase; (2) 17,20-lyase; (3) 3β-ol-dehydrogenase and Δ^5-isomerase and (4) 5α-reductase.

the hypothalamus and pituitary in the control of this early testicular function remains to be investigated. Fetal hypophysectomy from about 90 days (0.6 of term) leads to a decrease in gonadal growth in the sheep [395]. This is after the period of sex differentiation in this species.

Circulating plasma LH concentrations in chronically catheterised sheep fetuses are below the limits of sensitivity of assay (about 1 ng/ml) in the last third of gestation. Injection of 50 μg synthetic LH-RH to the fetus results in a

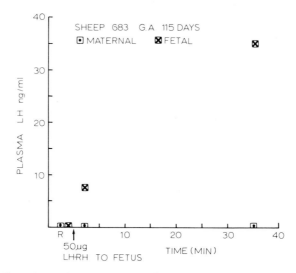

Fig. 3.3. The effect of injection of 50 μg LH-RH into a sheep fetus of 115 days gestation (Heap, Jenkin, Thomas, Jack & Nathanielsz, unpublished observations).

marked secretion of LH with plasma LH concentrations rising to 35 ng/ml (Fig. 3.3). These data demonstrate that the pituitary can respond to LH-RH in utero at this stage of development.

3.4 Role of androgenic steroids in development

The androgenic steroids are involved in two major processes in fetal life, sexual differentiation and parturition. The androgens involved in the processes of parturition are of fetal adrenal, placental and maternal origin. No differences in the delivery of single male or female fetuses have been noted and it is unlikely that the fetal testis plays an important role in parturition. Therefore, in this chapter we shall only concern ourselves with the role of androgens in sexual differentiation.

3.4.1 Embryological structures which give rise to the gonads and reproductive tract

(i) Development of the primitive testis has been described above. It should be recalled that the gland is comprised of several tissues of different origin. The gonad has two major functions: (a) to produce the germ cells, and (b) to synthesise and secrete various humoral agents; of these the best studied are the steroid hormones. In addition, the testis undoubtedly secretes other hormones of a protein nature, one of which, the Müllerian duct inhibitory factor, is discussed below.

(ii) The Wolffian (mesonephric) duct is the duct of the mesonephros, an organ which phylogenetically has a renal function but has no such function in mammals. Instead, this duct differentiates in males to form the epididymis, vas deferens and seminal vesicles (see Table 3.1).

(iii) The cephalic end of the Müllerian duct is produced by an invagination of coelomic endothelium and the caudal end is split off from the Wolffian duct. In the female it gives rise to the Fallopian tubes, the uterus and upper vagina. In the male it regresses under the influence of an unknown inhibitory testicular factor (Müllerian duct inhibitory factor).

(iv) The urogenital sinus which gives rise to the prostate and upper urethra in the male, and the lower third of the vagina in the female.

(v) The urogenital tubercle and urogenital swelling which give rise to the external genitalia.

The relationship of these structures is shown diagrammatically in Fig. 3.4.

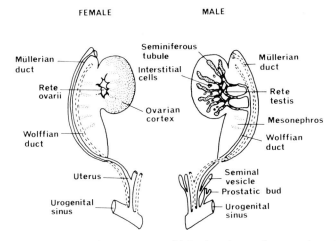

Fig. 3.4. The major embryonic structures which give rise to the gonads and accessory sex organs. Reproduced with permission from Jost [339].

3.4.2 Sex differentiation

The purpose of normal sexual differentiation is to produce adult members of the two sexes with gonads that produce functional gametes. These members of the species must display normal sexual behavioural patterns and possess reproductive tracts which enable fusion of the two forms of gametes at the time

Table 3.1.
Fate of different embryonic structures which give rise to the reproductive tract.

Embryonic structure	Definitive organs	Presence of testosterone-binding protein	Presence of 5α-reductase in tissue at time of differentiation	Probable regulator of differentiation
Wolffian duct	Epididymis Vas deferens Seminal vesicle	+ + +	− − −	Testosterone stimulates differentiation into definitive organs
Urogenital sinus	Prostate and upper portion of urethra in male Lower vagina and urethra in female	+	+	DHT stimulates differentiation in male
Urogenital swelling and urogenital tubercle	External genitalia	+	+	DHT causes male differentiation. In its absence the female phenotype occurs
Mullerian duct	Fallopian tubes Uterus Upper vagina	−	−	Regression produced in the male by a factor(s) from the testis (MDIF)

Table 3.2.
Development in different fetal situations. M = Male development pattern; F = Female development pattern.

Fetal structure	Normal genotypic male	Normal genotypic female	Male or female castrate[a]	Androgen-treated female	Genotypic male treated with cyproterone[b,c]
Wolffian duct	M	F	F	M	F
Mullerian duct	M	F	F	F	M
Urogenital sinus	M	F	F	M	F
Urogenital swelling and urogenital tubercle	M	F	F	M	F

[a] A similar situation exists in Turner's syndrome (genotype XO).
[b] A similar situation exists in the testicular feminisation syndrome.
[c] For report of the effect of administration of cyproterone to pregnant rats see [447].

of fertilization. In order to understand the role of fetal hormones in sexual development we must consider each level of differentiation.

Chromosomal sex

The classical view of chromosomal sex differentiation in mammals was that the heterogametic sex (XY) was male because of genes distributed on the Y chromosome and that the homogametic sex (XX) was female because she lacked the influence of the Y chromosome. The remaining pairs of chromosomes, the autosomes, were considered not to play any role in sex differentiation.

Certainly the Y chromosome is strongly gonad determining. If a normal Y chromosome is present then the primitive gonad forms as a testis, the medulla develops the sex cords and the cortex condenses to form the tunica. The male-determining factors are likely to be located on the short arm of the Y chromosome since there are rare cases in which the long arm of the Y chromosome only is present. These individuals are female. If the Y chromosome is absent, an ovary usually forms. There are exceptions to this as discussed below. In one sex chromosomal abnormality, Turner's syndrome, the genotype is XO, there is no Y chromosome and no second X chromosome. In this instance an ovary forms. It thus appears that one X is enough for differentiation of the ovary. In Turner's syndrome the ovary undergoes rapid atresia within months of birth and it has therefore been suggested that the function of the second X chromosome is to slow down the rate of ovarian atresia. Thus both X chromosomes are necessary for the normal development of the ovary. There are certain problems in this simplified explanation. For example, the XO mouse is fertile. However, the ovary only possesses half the number of ova compared with XX mice, and the XO mouse undergoes a premature menopause.

Recent evidence suggests that the Y chromosome cannot be responsible for all aspects of control of sex differentiation. Since the normal female possesses two X chromosomes, her genetic complement in this respect exceeds the male. The 'Lyon hypothesis' to explain how the organism compensates for this excess states that all but one of the X chromosomes in each cell undergoes condensation and forms heterochromatin (the Barr body) [269]. Heterochromatin is material which stains very darkly with basic dyes, whereas euchromatin stains more lightly, is more dispersed and is more metabolically active. In the normal female there is only one X chromosome to undergo heterochromatization (the genotype is XX). However, in certain abnormalities there may be several X chromosomes and suppression of X chromosomes is unaffected by the presence or absence of the Y chromosome, since it occurs as well in XXX abnormalities as in XXY.

Other features of sexual differentiation may also be independent of the Y chromosome. There can be autosomal determinants of sex differentiation. The function of the Y chromosome is to permit the expression of (or to switch on) certain male characteristics on the X chromosome. However, under certain abnormal conditions the determining genes are located on autosomes rather than on the Y chromosome. In any event, it is the genotype which determines the differentiation of the primitive gonad.

Gonads

The differentiation of the primitive gonad is thus controlled by the genotype. In all species studied so far the testis begins to differentiate before the ovary [339]. The control mechanism for this differentiation is not yet understood. The premature differentiation of the testis is compatible with the general rule presented above that male imprinting involves a positive overriding of the factors which will determine female patterns of development, whether they be endocrinological, structural or behavioural. It is also probably an evolutionary mechanism required by all mammals so that the male can develop in a maternal environment in which female hormones are present at high concentrations.

Administration of androgens to pregnant females early in development does not modify the development of the primitive gonad [521]. It has been proposed by Witschi that local inductor substances, produced in accordance with the genotype of the individual, determine whether the primitive gonadal cortex or medulla becomes functional or inactive. These putative substances have been called cortexenes and medullarenes but as yet there is little histological, biochemical or experimental evidence to support this idea in mammals. Parabiotic experiments, in which the male and female larvae of frogs and salamanders have been surgically conjoined, show that in sub-mammalian forms the testis can suppress ovarian development [586].

In mammals, an accident of nature permits the investigation of the possible effects of humoral substances on gonadal development. Occasionally, in a twin pregnancy, both fetuses of different sexual genotype may share a portion of the vascular supply to a portion of the placenta (Fig. 3.5). In this situation it has been demonstrated that differing degrees of degeneration and masculinisation of the ovary can occur [344]. The situation is variable according to the degree of vascular confluence and, of probable greater influence, the timing at which these vascular channels become effective. Such sex-reversed ovaries are capable of secreting testosterone [522]. The nature of the humoral factor is unknown. It is unlikely to be due to passage of germ cells from the testis of the male to its conjoined female twin since normal testicular development and androgen secretion can occur in the absence of germ cells. This is not the case in the ovary in which follicle formation depends on the presence of ova.

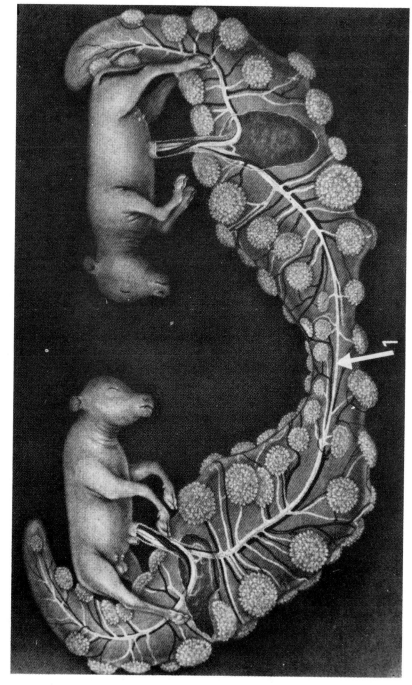

Fig. 3.5. Lillie's classical illustration of confluent, anastomosing blood vessels connecting the two placental circulations in twin calf fetuses. Modified and reproduced with permission from Lillie, J. Exp. Zool. 23, 371 (1917).

Gonadal hormones
The development of the capability to synthesise androgens and the effects of these hormones are considered below.

Reproductive ducts (internal genitalia)
The primitive structures which give rise to the reproductive tract have been described above and in Table 3.1. There are two factors which determine their differentiation in the male. Mullerian duct regression is not produced by androgen, whereas development and differentiation of the Wolffian duct is controlled by androgen.

Mullerian duct Elegant experiments by Jost and his colleagues have demonstrated that, whilst androgens do not cause Mullerian duct regression, the proximity of a testis to pieces of Mullerian duct in vitro will result in regression of the duct [468]. In the rat the Mullerian duct demonstrates a critical period of responsiveness. It appears that the testicular factor must be present from day $14\frac{1}{2}$ to day $17\frac{1}{2}$ to produce complete regression. Recent evidence suggests that the active principle has a protein structure [334]. This, as yet unidentified compound has been called Mullerian duct inhibitory factor (MDIF).

Wolffian duct Androgens administered to the mother or fetus of several species at a critical period of development will cause male differentiation of the Wolffian duct of genotypic females. In the sheep, the critical period appears to be between 20 and 40 days gestation [521]. This period corresponds well to the time at which androgens first appear in the fetal testis and the period at which differentiation of the Wolffian duct commences in vivo [29].

The mode of action of androgen has been most extensively studied in the rat and rabbit, particularly by Dr. J.D. Wilson [95, 96]. In the rat, with respect to the Wolffian duct, the active hormone is probably testosterone. As described below, in the case of the urogenital sinus (UGS) and urogenital tubercle (UGT) the active principle is probably the 5α-reduced androgen, 5α-dihydrotestosterone (DHT). In Wilson's experiments, administration of various natural and synthetic androgens to pregnant rats from day 14 resulted in masculinisation of the whole reproductive duct system of female embryos. Although DHT was active*, thus proving that this target tissue has receptors for DHT, the tissues of the Wolffian duct do not contain the 5α-reductase enzyme necessary to

---

* With respect to effects on sexual behaviour in the castrated adult, DHT is considerably less effective. This may be related to the fact that DHT cannot serve as an estrogen precursor.

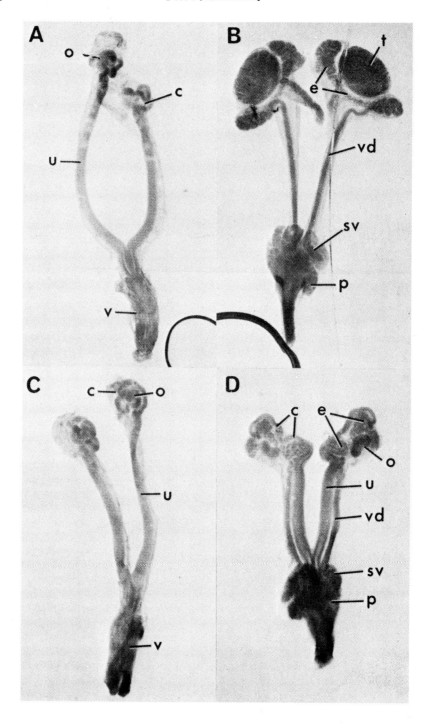

convert testosterone to DHT. 5β-DHT was inactive (Fig. 3.6) [510]. One point of interest from this work is that doses of DHT below those required for complete virilisation of the Wolffian duct system could produce unilateral differentiation, in which event the left side was always unaffected.

In summary, in the rat, rabbit and sheep, administration of androgen to female fetuses will cause virilisation of the Wolffian duct system if given at a critical period. This period coincides with the time at which testicular androgen synthesis can be demonstrated. Specific receptors for androgen have been demonstrated in the Wolffian duct system [584, 587].

Lower reproductive tract and external genitalia (phenotype)
Experimental investigation of control of differentiation of the UGS and UGT has followed similar lines to those described above for the Wolffian duct. With these tissues as well, we observe the rule that in the castrate of both sexes, the female pattern emerges. The factor which controls male differentiation is androgen [447]. As shown in Table 3.1, these tissues do possess 5α-reductase in contrast to those of the Wolffian duct system. In several systems DHT is a far more potent androgen than testosterone in both fetus and adult [584]. Also, congenital absence of 5α-reductase activity, as in the human testicular feminisation syndrome, results in the inability of the testis to masculinise the lower reproductive tract. It should be noted that the action of testosterone on the Wolffian duct is not affected and that the Mullerian ducts regress as expected since their regression is not controlled by androgen.

Neural and behavioural differentiation
The behavioural characteristics associated with the male of the species are complex patterns influenced by many different factors. It has, however, been demonstrated that the basic neural differentiation follows the same primary law of sexual differentiation stated before. For male behaviour to be demonstrated, androgen must act on the developing brain at a critical point in develop-

◀ Fig. 3.6. The effect of DHT and 5β-DHT on the urogenital tract of the newborn female rat. Pregnant rats were given either triolein or 16 mg of hormone dissolved in triolein each day by subcutaneous injection from days 14 to 21 of gestation, and newborn rats from these animals were given 0.64 mg each on days 1 and 3 after birth and were killed and dissected on day 4. (A) Control female urogenital tract. (B) Control male urogenital tract. (C) Female urogenital tract following administration of 5β-DHT. (D) Female urogenital tract following administration of DHT. Key: o, ovary; u, uterus; c, coils of oviduct; v, vagina; t, testis; e, epididymis; vd, vas deferens; sv, seminal vesicle; p, prostate. Reproduced with permission from Schultz and Wilson [510].

ment. In the rat this period is the first five days of neonatal life. Exposure of the neonate to androgen at this critical period results in modification of the inherent rhythmicity of the hypothalamus which would otherwise produce cyclical release of gonadotropins from the pituitary, thus giving rise to the well known female reproductive cycles [27, 89, 201, 270, 542]. Other hormones may synergise in producing this effect [366].

This period of plasticity is terminated in utero in several species born at a more mature phase of development [527]. Thus, the administration of testosterone proprionate* to the pregnant guinea pig for short periods (between day 33 and 37 of pregnancy) not only produces masculinisation of the lower reproductive tract of any female fetuses but also affects the rhythmicity of ovarian function when these fetuses reach puberty to such an extent as to prevent mating. Although there are certain differences; these results suggest certain similarities to the observations of the effect of testosterone injection into the neonatal rat [90]. Similarly, administration of testosterone implants to pregnant ewes before day 60 of pregnancy produces male behaviour in the fetuses when they reach adult life. As well as the behavioural changes, it can be demonstrated that the normal responses of the adult female sheep to estrogen administration are absent in the ewes which were exposed to androgen at this critical phase [521].

At the present time, the studies referred to above have yielded considerable insight into the development and differentiation of sex differences in the human. An understanding of normal sexual differentiation is an essential prerequisite to the unravelling of the pathogenesis of the multitude of abnormalities of sexual development in man. Several questions remain to be answered by further work.

---

* Wilson observed that testosterone administered to the pregnant rat caused resorption of the fetuses. This does not occur in the guinea pig.

CHAPTER 4

General features of the development and function of the fetal hypothalamo–hypophysial–portal system

Fetal hypothalamic control of adenohypophysial function has been clearly demonstrated by experiments in the fetal sheep and rat. In the sheep, Liggins interrupted the vascular and axonal connections between the hypothalamus and pituitary by sectioning the pituitary stalk and then inserting a silicone rubber plate between them to prevent regeneration. This type of surgery can be performed anytime after about 85 days of gestation (term is approximately 150 days). Vascular catheters can be placed in fetal blood vessels at the time of stalk section. The fetal blood vessels usually catheterised are the jugular vein, carotid artery or the dorsal aorta or inferior vena cava with catheters advanced towards the heart after insertion into blood vessels in the fetal leg. Fetal blood samples can then be taken for measurement of plasma concentrations of pituitary hormones. In this way it is possible to assess the effect of isolation of the pituitary from the hypothalamus. Stalk section preparations may be maintained in utero for up to 30 days or more. If the section is successful, parturition does not occur and delivery of the fetus has to be effected by caesarian section (see Chapter 12). In the sheep the pituitary is supplied by arterial blood from the Circle of Willis as well as blood from the pituitary portal system. If the surgery for stalk section has compromised the arterial supply as well as sectioning the portal vessels, infarction of the pituitary will occur [389]. At the termination of the experiment, careful histological examination of the pituitary will determine to what extent, if any, the pituitary has undergone necrosis. Specific studies with stalk-sectioned lambs will be discussed when considering individual systems.

In general, assessment of the onset of the early stages of development of the portal system in different species has been attempted by two techniques; firstly, vascular perfusion studies with markers such as Indian ink; secondly, surgical attempts to isolate the various levels of activity which are shown diagrammatically for the thyroid axis in Fig. 4.1.

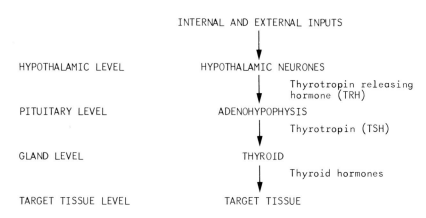

Fig. 4.1. Simplified diagrammatic representation of the major levels of endocrine activity in the thyroid axis.

4.1 Perfusion experiments

Early attempts to perfuse the vascular bed in the region of the hypothalamus and pituitary with Indian ink particles were carried out in the newborn rat pup. These experiments led to the conclusion that the primary plexus could only be physically present as a functional structure at day 5 postnatally or soon after [102, 245]. It is likely that the Indian ink particle size generally used (approx. 50 μm) limits the usefulness of this technique, rendering it incapable of demonstrating any very small capillary channels as they develop. In this respect it is perhaps worth recalling that the negative result to an experiment is always difficult to interpret. It is preferable to make only one conclusion from an experiment which yields a negative result – namely that 'using the experimental techniques described, it was impossible to demonstrate the particular phenomenon'. A negative result does not prove the non-existence of the phenomenon investigated.

When markers of smaller dimensions such as radioactively labelled colloid particles or red cells were used, a good, well-established portal flow could be demonstrated at 4 days of postnatal age in the rat. This was the earliest age investigated [221]. This observation lends weight to the idea that the problem with investigation of the development of the portal system in the rat is related to the respective size of the vessels and the particles used to assess the extent of blood flow. Further evidence for this view is provided by the work of Campbell [102] who was able to demonstrate the primary plexus of the portal system in the rabbit at 17 days gestational age (term = 32 days). The secondary plexus was visible at 18 days. In contrast to this ability to show the components of the portal system in the fetal rabbit, he was only able to confirm the earlier work in the rat and demonstrate the portal system at about the 5th postnatal day.

Use of the electronmicroscope for detailed study of this region in the fetal and neonatal rat may be a better method of assessing the development of these structures. Various forms of vesicle have been demonstrated in the nerve fibres of the median eminence between day 16 of gestation and the first day of neonatal life in the rat [205, 529]. Fenestrated capillaries were present in this region but they had not thoroughly invaded the area.

Summarising the structural studies, it appears that signs of competence of the neurosecretory link are available in the rat at the earliest ages studied in neonatal life. In the fetal rabbit a well-developed portal system can be shown by 18 days gestation. In the fetal sheep it is clear that there is a functional link in the latter part of pregnancy.

4.2 Functional studies following ablative surgery

Very little data exists regarding the timing of onset and extent of activity of hypothalamic hypophysiotropic hormones. However, some evidence regarding their function in the latter part of pregnancy is available. Pioneer studies on the role of the interrelationship of the fetal hypothalamus and pituitary were begun by Jost [335, 340, 341] in the rat and the rabbit. The pre-eminence of the fetal sheep for the study of this system was demonstrated by the early studies of Liggins.

4.2.1 The rat

There are three basic preparations available for the study of the level of activity of the various endocrine axes involving the hypothalamus (see Fig. 4.2). These preparations have been developed primarily as the result of the work of Jost and his colleagues.

(i) The intact fetus: all endocrine structures in the axis are present.

(ii) The encephalectomised fetus: the whole brain is removed but the pituitary gland is left in situ.

(iii) The decapitated fetus: section is low enough to remove all the brain, including the pituitary but the thyroid gland is left in position.

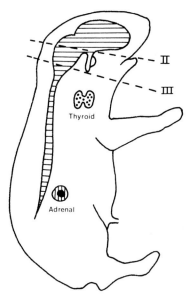

Fig. 4.2. Diagrammatic representation of various fetal rat preparations used to study the development of endocrine activity. We may call Preparation I the intact animal. Preparation II shows the level of encephalectomy, involving removal of most of the brain, including the hypothalamus but leaving the pituitary intact. Preparation III shows the effect of decapitation which involves removal of the pituitary gland as well. It should be noted that the level of section is above the thyroid gland. The adrenal gland is also shown.

As described above, there is some dispute regarding the conclusions of the various investigations of structural links between the hypothalamus and pituitary in the rat fetus. However, firmer conclusions can be drawn from the experiments to be considered individually for GH, TSH, ACTH, and the gonadotropins based on observations of various end points of endocrine function. The available evidence suggests that the fetal rat hypothalamus is capable of exerting a controlling influence on the pituitary. The questions to be asked are 'When does this influence first manifest itself under physiological conditions?' And 'To what extent is the connection active in different situations during fetal life?'.

It has been suggested that differentiation of the adenohypophysis may await the establishment of a functional hypothalamo–hypophysial link [263]. Such a possibility is in keeping with the ideas discussed in Chapter 9 that tropic hormones may play a role during development by inducing the differentiation of their own receptors in their target endocrine cells. A different view is held by some workers who state that histological differentiation of the pituitary precedes any functional connection between the hypothalamus and pituitary. If this is true, then clearly the hypothalamic releasing hormones cannot affect the initial differentiation of the pituitary. This idea is further developed by some workers who hold that gradual evolution of hierarchical control has occurred in each hypothalamo–pituitary axis. According to this view, adenohypophysial function existed before it came under hypothalamic control. If this concept is correct then it is unlikely that the hypothalamus exerts any developmental control over the pituitary. An additional factor which requires consideration is the possible existence and role of hypothalamic inhibitory hormones. Clearly further work is required in this area and it would be useful to have direct measurements of circulating and tissue hormone concentrations to relate to histological observations. Fetal rats removed from the uterus and exposed to ether stress showed evidence of increased hypothalamic CRF content and release [288]. This is one index of ability of the hypothalamus to function at a maturity earlier than that at which birth would normally have occurred, although the animal was no longer in an in vivo situation.

4.2.2 The sheep

In this species selective stalk section and hypophysectomy can be performed. The method of stalk section used by Liggins has been described above. The technique used by Thorburn would appear to be the same but the preparation produced is different (see below). However, the number of stalk-sectioned sheep fetuses from which information is available in the scientific literature is 17. Most of these were in the early years in which chronic fetal sheep preparations were performed and it is only in a few of these animals that fetal endocrine changes have been followed sequentially. As measured by plasma concentrations, two reports state that pituitary secretion of GH is greatly depressed after stalk section whereas the effect on TSH is variable [554, 577]. When considering data from stalk-section experiments, two points should be borne in mind. Firstly, the extent of stalk section and, secondly, the competence of the adenohypophysis to respond to hypothalamic releasing hormones to assess whether any pituitary damage has occurred during surgery. In view of the various changes in feedback and the very real possibility of changes of levels of activity at different stages of gestation, attention should be paid to the various indices which might be used to assess these important criteria of each preparation.

As mentioned earlier, one crucial problem to bear in mind when assessing the results from fetal stalk-section experiments is that, as with stalk section in the adult sheep, this procedure is not without the risk of subsequent pituitary infarction [5, 153]. This important question is discussed by Liggins and Thorburn in a recent CIBA Foundation Symposium [389, p. 170; 552, p. 202]. Continued fetal growth was observed in lambs stalk-sectioned by one technique, whilst with a different technique used by Thorburn, in which a lower section was made with a resulting pituitary infarct, the fetuses did not grow. There are two possible explanations for this apparent difference. Firstly, high stalk section may retain a small median eminence effect on the pituitary which results in continued growth. Alternatively, the infarct produced by the lower section almost certainly decreases pituitary reserve secretion. Hence growth retardation may result from a state of partial hypophysectomy. Since the preparations in which the fetuses continued to grow were subjected to detailed post-mortem histological examination involving serial sections examined microscopically, the second alternative seems more likely at the present time.

I would suggest the following methods directed towards firmer interpretation of the results obtained from stalk section and hypophysectomy experiments in the sheep fetus.

(1) Simultaneous measurement of more than one pituitary hormone will avoid complications of interpretation of possibilities of feedback with its attendant time- and concentration-dependence.

(2) Measurement of pituitary reserve secretory capacity by some provocative test. We have used 50 μg thyrotropin-releasing hormone (TRH) administered intravascularly to the fetus (Fig. 4.3). This should enable an assessment of

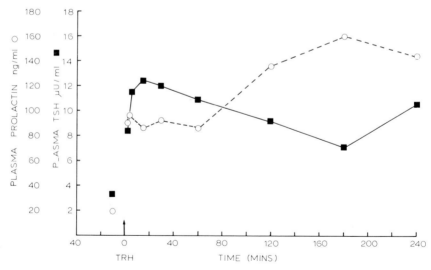

Fig. 4.3. The effect of intravenous injection of 50 μg TRH into a lamb fetus of 132 days gestation. Reproduced with permission from Biol. Neonate 26, 109-116, 1975.

viability and function of the pituitary. Another possible test which may prove useful is hypoglycemic challenge to the pituitary adrenal axis – not without its dangers (!). As mentioned in Chapter 3 we have also obtained LH release from the fetal pituitary in response to LH-RH (see Fig. 3.4). The practical significance of these provocative tests is that the results of radioimmunoassay measurement of prolactin or LH can be available within one or two days of the test. It is then possible to assess the state of the preparation.

(3) Careful macroscopic and microscopic examination should always be carried out at the termination of the experiment.

4.3 Setting of level of hypothalamo–hypophysial activity during fetal life

The final level of activity in any of the hypothalamo–hypophysial systems will depend on the interaction of several inputs (see Fig. 4.1). It is now clear that the developing brain has certain critical phases of development at which future activity can be permanently affected. Three examples are mentioned in this book, (i) the conditioning of sexual rhythmicity, (p. 43), (ii) the effect of glucocorticoids on the development of nyctohemeral, or circadian, rhythms in glucocorticoid secretion (p. 133) and (iii) the actions of thyroid hormones. Most of the work with thyroid hormones has been performed in the neonatal period using the rat. Excess thyroid hormones administered to newborn rats in the first two weeks of life produce permanent impairment of growth and a resetting of the level of activity of the thyroid axis. Thus high doses of thyroid hormones produce permanent reduction of pituitary and plasma TSH concentrations [34, 35, 65]. Short periods of neonatal hypothyroidism can also result in irreversible alteration of the feedback system [34]. Other hormones may also affect the level of function of the thyroid axis. Administration of 2 mg testosterone proprionate to rats at 36–48 hours of age results in a depressed thyroid secretion rate when these rats reach 120 days of age [374].

T_4 plays a role in the mechanisms whereby the body controls its basal metabolic rate. In addition, T_4 secretion increases at times when increased heat production is required. It appears that the maturation of the temperature regulation mechanisms which are sited predominantly in the hypothalamus, is also affected by T_4. The maturation of these systems is retarded and incomplete in hypothyroid rats [267]. It has been proposed that a change in sensitivity of the hypothalamus to the temperature of the blood perfusing it during the last few days of gestation plays an important role in the mechanisms which initiate delivery [554a]. Fetal blood is at 0.5–0.8°C higher temperature than the neonatal or adult core temperature. According to this theory, about 10 days before delivery the hypothalamic temperature regulatory centres are reset to a new, lower balance point for all the various inputs. As a result the fetus initially perceives that its temperature is above the new set point. One of the mechanisms whereby heat production is decreased is to decrease T_4 secretion. This attractive hypothesis of maturation of a feedback system is discussed in Chapter 5.

CHAPTER 5

The fetal thyroid. I. General features and experimental studies in ruminants

5.1 Developmental aspects

5.1.1 Evolution of the thyroid gland

The larval form of the lamprey, a primitive cyclostome living in fresh water, and the adult forms of certain protochordates (forerunners of the vertebrates) such as Amphioxus, have a clearly defined glandular structure in the floor of the pharynx. This is called the subpharyngeal gland in the cyclostomes or the endostyle in Amphioxus. The apparent purpose of this structure is to produce mucus from typical mucus-producing cells. The food is mixed with this mucus. In these animals, other cells which can trap iodide and produce iodinated tyrosine residues are also present within this structure. These iodide-trapping cells are quite distinct from the mucus cells.

There are two major theories regarding the evolution of thyroxine (T_4) secreting cells into the compact glandular structure characteristic of mammals. One idea is that the iodinated amino acids play a hormonal role even in the primitive forms described above. The iodinated amino acids are thought to reach the blood stream after exocrine secretion from the pharyngeal cells and eventual absorption from the gut. A parallel may be drawn with the enterohepatic circulation of T_4 in adult mammals during which the sulphate and glucuronide conjugates of T_4 pass into the gut via the bile duct. This conjugated T_4 is eventually hydrolysed and absorbed from the intestine (see Fig. 5.1). According to this theory of the evolution of the thyroid gland, the juxtaposition of the mucus-producing cells and the iodide-trapping cells in the pharyngeal region was purely coincidental. The mucus cells were eventually left behind when the T_4-producing cells invaginated, losing their contact with the pharynx, and gave rise to the adult mammalian thyroid gland.

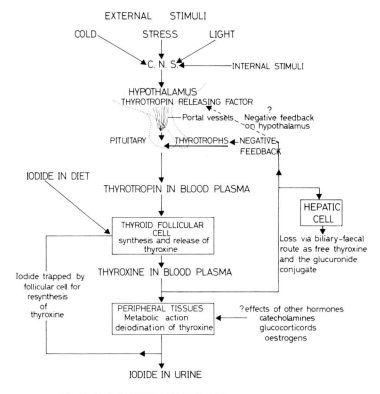

Fig. 5.1. Some of the important features in the distribution and regulation of iodide and thyroxine metabolism in the body.

The second theory is that, with the assumption of a freshwater existence, the mammalian organism required an iodide-trapping mechanism. These subpharyngeal cells were well-suited to fulfil this role. These two theories should be borne in mind when a consideration is undertaken of the role of the thyroid in embryonic and very early fetal development. They have a particular relevance to the suggestion that elemental iodine may play a role in development which is distinct from the activity of the iodothyronines.

5.1.2 Embryological maturation of the thyroid gland

Structural development
The evolution of the mammalian thyroid gland is recapitulated in the developing fetus at both the structural and biochemical level. This chapter contains

a considerable amount of data on the kinetics of thyroid function in the fetal mammal which has been obtained from chronically catheterised fetal preparations. Much of the earlier work on the development of fetal thyroid function was directed towards investigation of the structural differentiation of the gland and assessment of the stages at which thyroglobulin and the different iodinated compounds could be isolated from the fetal gland.

The first sign of differentiation of the thyroid gland is the appearance of the thyroglossal duct as an outpouching from the floor of the developing buccal cavity. This occurs at a point which coincides with the junction of the anterior two-thirds and posterior third of the tongue. Eventually, continuity with buccal cavity is lost as the gland is pulled caudally whilst the neck elongates. During this phase the developing thyroid gland is invaded by cells from the ultimobrancial body (the fifth pharyngeal pouch). These cells are the parafollicular cells which secrete thyrocalcitonin*.

After the thyroid has reached its adult position in the neck, histological differentiation occurs. Initially the gland is composed of a solid mass of cells with no follicular structure and no colloid – this is called the pre-colloid stage. Eventually small intracellular cannaliculi appear in the cells and these coalesce to form the central colloid space in each follicle. This is followed by a phase of colloid secretion and follicular growth. These stages have been shown to occur in a similar progression in the human, sheep, mouse, rat and pig [37, 70, 202, 208, 248, 291, 325, 517, 518].

Functional development
Studies of the functional capacity of the fetal thyroid gland have been conducted in relation to the various steps in T_4 synthesis by the follicular cells. It appears that the ability of the gland to trap iodide precedes the onset of synthesis of iodinated amino acids. In the fetal sheep this first occurs at about 50 days gestation and in the human at about 84 days. Development of the ability to

---

* Thyrocalcitonin is a polypeptide hormone whose action is to lower plasma calcium concentration. The major mechanism of action of the hormone is to increase the deposition of calcium in bone. It probably plays a very important role in bone development in utero. At the present time insufficient work has been carried out on the function of thyrocalcitonin in fetal life to merit consideration in detail here. Radioimmunoassay systems for thyrocalcitonin and the hypercalcemia-producing, bone-mobilising hormone, parathormone, are now available in the human, sheep and ox [13]. This methodology, together with a better understanding of the different forms in which plasma calcium circulates, should be an adequate stimulus to a detailed investigation of the control of calcium in the fetus.

synthesise T_4 follows the gradual appearance of the individual steps in the biosynthetic pathway. Iodinated amino acids can be extracted from the fetal sheep thyroid at 70 days gestation and the human fetal thyroid at about 100 days. This biosynthetic pathway for T_4 synthesis is well-defined. The developing thyroid cell should therefore prove a good model tissue for biochemists with an interest in intracellular regulatory mechanisms. It should be possible to investigate the influence of different steps in biochemical pathways on the development and differentiation of subsequent steps in the same metabolic pathway*. With the use of sophisticated histological techniques small amounts of thyroglobulin can be demonstrated within the cells of the developing thyroid before the appearance of a follicular space. It is therefore sometimes stated that biochemical differentiation precedes histological differentiation. This seemingly profound comment is obviously true, since histological differentiation merely represents biochemical differentiation advanced to a sufficient degree to be visible as a structural change.

Investigations of thyroid function in early fetal life have generally involved the administration of radioactively labelled iodide (^{125}I-Na or ^{131}I-Na) to mother or fetus. At different time intervals following the administration of the tracer, estimations are made of either total thyroidal radioactivity or the radioactivity that has been incorporated into the various thyroid compounds after preliminary separation of thyroidal and blood components into various iodinated compounds using chromatographic techniques. Whilst these methods have given useful information on the biochemical differentiation of the fetal thyroid gland, a degree of caution should be exercised in any attempt to draw too firm conclusions from this type of experiment. Failure to demonstrate the presence of an intermediate or even a final product within an endocrine gland does not necessarily prove the inability of the gland to synthesise the compound under all conditions, even at the stage of development investigated. Too often very little attention has been paid to the time-course of uptake, incorporation and release of radioactively labelled precursors. In addition, the extraction of the gland in such experiments has been undertaken at very arbitrary and often unsuitable time intervals after the administration of the tracer. It is also very necessary to make observations at different periods of development when undertaking this type of study. These points are considered again in Chapters 10 and 11 with relation to steroid synthesis. Several additional factors will affect the experiment when the tracer is administered to the mother. In particular, it is necessary to consider the effect of placental transport of iodide.

---

* Inherited defects in the biosynthesis of thyroid hormones have been reported in the sheep [481].

In summary, it can be seen that at a very early stage in development the fetal thyroid can produce T_4 and triiodothyronine (T_3) in all species studied. The thyroid gland content ratio for $T_4:T_3$ is similar in fetal and adult sheep and fetal and adult humans [126, 209].

Several questions follow from the observations described above. Firstly, is there a phase before the assumption of fetal thyroid secretory competence when the fetus is dependent on T_4 (or a different molecule with a similar function) which it obtains from a source other than its own thyroid? Does elemental iodine have a role to play in the early stages? Secondly, what is the level of fetal thyroid activity at different stages of development? Are there any conditions in which maternal thyroid activity augments fetal thyroid secretion? Finally, it must be noted that it is impossible to consider fetal thyroid action in isolation. The thyroid gland is one link in the hypothalamo–hypophysial–thyroid–target tissue axis (Fig. 5.1.).

5.2 The hypothalamo–hypophysial–thyroid–target tissue axis in the adult mammal

5.2.1 Hypothalamus and adenohypophysis

After removal of the pituitary gland, the level of thyroid function is depressed to less than 5% of normal in adult and fetal mammals [544, 548, 549]. This depression of activity results from the fact that one of the basophilic cell series in the adenohypophysis produces a glycoprotein hormone, thyrotropin or thyroid-stimulating hormone (TSH). TSH has several actions on the thyroid gland, affecting almost all the steps involved in the synthesis and secretion of T_4 [555]. The secretion of TSH is in turn controlled by a tripeptide (ptyro-glutamyl-histidyl-prolineamide, or thyrotropin-releasing hormone (TRH) secreted by nerve cells whose nuclei are thought to lie in the arcuate nucleus of the hypothalamus. These neurones send out nerve fibres terminating on the primary plexus of the portal system which forms a vascular link between the hypothalamus and adenohypophysis. These are the Type I neurosecretory neurones shown in Fig. 1.4. These neurones act as a final common path for external and internal stimuli which change the rate of release of TRH and hence control the rate of TSH secretion by the adenohypophysis (see Fig. 4.1). TRH has been synthesised and is available commercially. When administered intravenously, TRH can be used as a stimulus to test the ability of the pituitary to release TSH (see Fig. 4.3).

In the adult mammal it is necessary to produce bilateral hypothalamic lesions to prevent the transmission of hypothalamic influences to the pituitary secretion

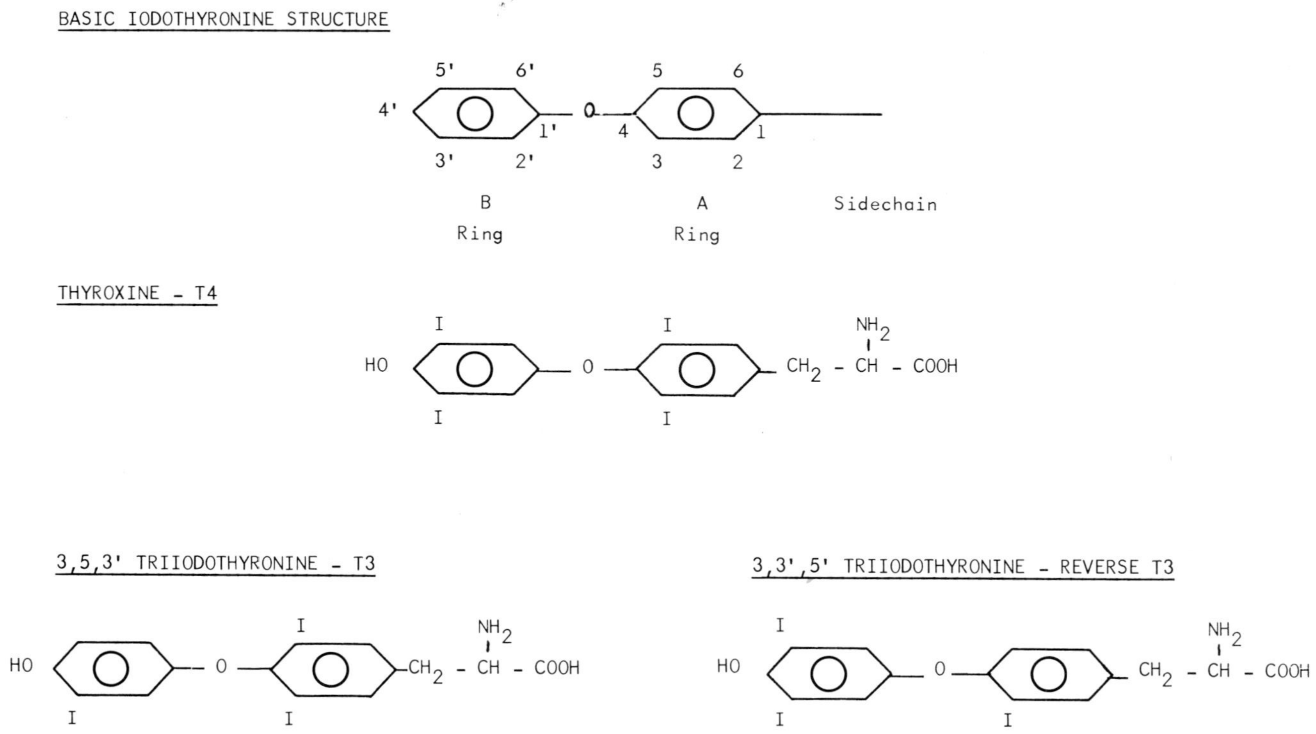

Fig. 5.2. Structures of the various iodothyronines considered in this chapter.

of TSH [555]. This experimental observation reflects the presence of paired inputs to the various bilaterally placed hypothalamic nuclei. The nature and extent of inputs to the adult hypothalamus have received considerable investigation in relation to each of the controlling systems affecting the release of adenohypophysial hormones. One of the major stimuli to increased activity appears to be cold exposure though the relative efficiency of acute and chronic cold exposure in releasing TRH, TSH and T_4 may vary between species and also at different stages of life even within a species. The general principles of neurosecretion and hierarchical control have been considered in Chapter 1. Fig. 5.1 demonstrates diagrammatically the various links in the thyroid system of the adult animal

5.2.2 The hormones of the thyroid gland

The structures of the two active hormones secreted by the thyroid gland are shown in Fig. 5.2. T_4 and T_3 released from the thyroid gland are carried in the blood in a reversible equilibrium: some of the circulating hormone is bound to plasma proteins and a portion is free or diffusible (see Chapter 2). The actual nature of the binding proteins varies from species to species. There are also differences in the proportion of diffusible hormone. The diffusible T_4 is generally considered to be the physiologically active portion. The general principles of plasma protein binding of thyroid hormones are the same in all mammalian species. The physiological consequences of protein binding on hormone function have been the subject of several recent reviews [536, 593, 459]. In general the concentrations of free T_4 in fetal plasma are greater than in maternal plasma.

Each target tissue of the endocrine axis is exposed to the free hormone which diffuses from the plasma into the extracellular tissue spaces. Hormone receptors on the cell surface bind the hormone to the target tissue. T_4 binding has been demonstrated in various tissues, particularly the liver and kidney. However, the binding of T_3 may be more important since the avidity of the liver and kidney nuclear binding sites which are specific and of limited capacity, is greater for T_3 than T_4. The adenohypophysis also has specific T_3 binding sites and these may be important in regulating the negative feedback of thyroid hormones on TSH secretion.

5.2.3 The role of T_3 and structure–activity relationships for iodothyronines

T_3 and T_4 are both effective in reversing hypothyroidism following thyroidectomy. Although there is some debate as to the extent of their different potencies, it is now clear that T_3 is more effective than T_4. Inhibition of peripheral monodeiodination of $T_4 \rightarrow T_3$ with thiouracil drugs will interfere with the action of T_4. From calculations of the appearance of T_3 in the plasma after the injection of T_4 into thyroidectomised animals, it appears that as much as 30% of the T_4 secreted by the thyroid is converted to T_3 in the adult. Since T_3 is about 3 times as potent as T_4 in the adult human, T_3 can probably account for all the activity of T_4 [535]. Deiodination of T_4 at position 5' in the outer B ring gives rise to $3,5,3'$-T_3, triiodothyronine (Fig. 5.2). Deiodination at the 5 position in the A (inner ring) will produce $3',5',3$-T_3 (reverse T_3; rT_3). rT_3 may not be biologically inactive as usually stated but may be an antagonist of T_4 (see below).

Using $[3:5\ ^{131}I_2, 3':5'\ ^{124}I_2]$thyroxine it is possible to follow separately any deiodination in the outer B ring, here labelled with ^{131}I, and the inner A ring, labelled with ^{124}I. It has been demonstrated both in vivo and in vitro that deiodination does occur in the inner A ring but apparently at a slower rate than in the outer B ring. This interesting report of experimental studies on monodeiodination discusses the various theories regarding structure–activity relationship among the iodothyronines. It appears that iodine is necessary on both sites (3 and 5) on the A ring for iodothyronines to retain their activity. The author makes a point of fundamental importance when discussing this problem. 'The finding that 3'-hydroxy-3:5-diiodothyronine is physiologically inactive does not necessarily conflict with this view [regarding the A ring iodine] because this substance is labile and may never reach the target organs intact. The active form of the hormone may well be a labile compound which is formed in situ within the cell and which shows no physiological activity when presented to tissue extracellularly'. [473].

The mode of action of T_4 and T_3 at the cellular level has for long been a matter of controversy. 'After nearly 30 years of intensive research we do not yet know how to explain the physiological actions of thyroid hormones in terms of molecular mechanisms', writes one recent reviewer [546].

One of the major problems with in vitro work using thyroid hormones* is that the experiments have generally been carried out with concentrations of free T_4 of around 10^{-5} M whereas the physiological concentrations of free hormone in plasma are of the order of 10^{-10} M [249]. For example in one study which demonstrated the inhibition of oxidative phosphorylation by T_4, the hormone was used at a free T_4 concentration of 1.3×10^{-5} M [379].

---

* Throughout this chapter, reference is made to 'thyroid hormones' when it is not intended to draw any distinction between T_4 and T_3 but simply to refer to the active agents of the thyroid gland.

In most tissues, thyroid hormones will increase the metabolic rate and tissue oxygen utilization. An exception is neural tissue in which oxygen consumption is not affected. Studies with liver cells show that thyroid hormones increase the population of ribosomes and enhance both mRNA and rRNA synthesis. They synergise with other hormones such as growth hormone and testosterone when studied in vitro [581]. The various changes observed are those which would be expected to accompany increased protein synthesis. Recently it has been suggested that changes in the synthesis of the enzyme Na^+, K^+-ATPase is the most fundamental effect of thyroid hormones. This enzyme is responsible for the maintenance of intracellular ionic composition to produce the correct cellular resting membrane potential. This potential difference is fundamental to the activity of all cells – especially neurones and secretory cells. Energy expended in these ionic pumping mechanisms has been calculated to account for about 50% of the basal metabolic rate (BMR). The absence of an effect of thyroid hormones on oxygen consumption by neural tissue probably reflects the lack of action of thyroid hormones on the Na^+, K^+-ATPase in this tissue. This may be of extreme importance since the resting potential of the nerve fibre must be kept within very critical limits [187].

5.3 Experimental observations on fetal thyroid function in different species

5.3.1 Sheep

Fetal hypothalamo–hypophysial–thyroid autonomy
The sheep is the species in which the most extensive investigations of the activity of the thyroid axis in individual fetuses have been carried out. This is true of the function of all the different levels of the axis and at most stages of gestation. Data of measurements of the various important hormones of the thyroid axis are given in Table 5.1. In summary, between 50 and 140 days gestation the level of activity of the fetal hypothalamo–hypophysial–thyroid axis appears to be considerably greater than in the adult. Plasma TSH concentrations in the fetus are 3–5 times those of the mother. Turnover of T_4 in the fetal compartment also exceeds that in the maternal compartment by a factor of 3–4. In contrast, fetal plasma T_3 concentrations are low, generally less than 150 pg/ml and rT_3 concentrations are high (around 5 ng/ml).

It is necessary to state at an early stage of our consideration of the fetal thyroid axis that, after 80 days gestation, the fetal sheep has been shown to be an independent, autonomous unit at all levels of the axis at all ages studied. In the latter half of gestation the fetus obtains iodide via the placenta which can

Table 5.1.
Various values of plasma concentrations for different components of the thyroid axis in human and sheep. Data taken from [125, 193, 306, 443, 445]. MCR = metabolic clearance rate, expressed in l/m² /day.

	Gestational age (days)	Thyroxine (T₄)					TSH (μU/ml)	Triiodothyronine (T₃)					Reverse triiodothyronine (rT₃)					
		Total T₄ (ng/ml)	% Free T₄	Total free T₄ (pg/ml)	MCR T₄	T₄ production (μg/m²/day)		Total T₃ (ng/ml)	% Free T₃	Total free T₃ (pg/ml)	MCR T₃	T₃ production (μg/m²/day)	Total rT₃ (ng/ml)	% Free rT₃	Total free rT₃	MCR rT₃	rT₃ production (μg/m²/day)
SHEEP:																		
Fetus	80–140	91	0.10	9.1	3.85	335	2.8*	0.016	0.63	0.10	80.2	27	3.79	0.41	15.53	18.8	101.8
Ewe	80–140	66	0.07	4.6	2.52	146	2.4	0.077	0.46	0.35	42.2	28	0.96	0.34	3.26	73.7	38.0
HUMAN:																		
Fetus	77–126	26	0.07	18.5			2.4										
Mother	77–126	129	0.02	29.5			4.2										
Fetus	154–238	72	0.04	24.9			9.6										
Mother	154–238	122	0.02	28.2			3.8										
Fetus and neonatal cord blood	226–280	112	0.03	29.0			8.9	0.25					1.51				
Mother	226–280	115	0.02	23.0			4.3	1.26					0.41				

* ng/ml.

concentrate iodide to a fetal: maternal ratio of 8 : 1 [247, 443]. The use of radioactively labelled T_4 and T_3 injected into both fetus and mother have shown that the placenta is relatively impermeable to T_4 and T_3 [99, 134, 443, 496, 497]. Fetal thyroidectomy results in an increase in fetal plasma TSH concentrations [554].

TRH administered to the ewe in large doses which increase TSH, T_4 and T_3 in the maternal circulation does not result in any change in thyroid axis hormones across the placenta in the fetus [446, 551]. Similarly fetal TRH injection does not affect maternal hormone concentrations. The amounts of TRH injected into the ewe (200 µg) and fetus (50 µg) via the peripheral vasculature produce very good pituitary responses. In the physiological situation TRH is released locally into the portal system and much lower concentrations would be found circulating in the periphery. These experiments therefore expose the placenta to supramaximal plasma TRH concentrations. In spite of this, no effect on the thyroid axis manifests itself on the opposite side of the placenta. We may conclude that there is no significant placental transport of TRH, TSH, T_4 or T_3 in either direction. No data are yet available regarding the changes in rT_3 when the fetal thyroid axis is stimulated, similarly the placental permeability to rT_3 is unknown though the F : M concentration gradient of greater than 20 : 1 suggests that the placenta is impermeable to rT_3.

Hypophysectomy of the fetal sheep is followed by a fall in fetal plasma TSH concentrations to undetectable levels within 2 days and plasma T_4 concentrations in 4–5 days [554]. These observations again demonstrate the autonomy of the fetal thyroid axis. The results of the same group of workers with 5 stalk-sectioned preparations are shown in Fig. 5.3. It is clear that several different TSH and T_4 pictures occur. These differences are probably related to the varying extent of pituitary infarction described with this stalk-section technique (p. 49). The presence of some degree of infarction makes it difficult to assess the activity of the feedback system in these preparations isolated from hypothalamic activity. It is for precisely this reason that the suggestion is made in this book that future experiments on fetal hypothalamic and pituitary activity in the sheep should monitor more than one pituitary hormone to obtain an indication of the extent to which pituitary damage may explain the observations. Detailed histological examination of the material is also necessary.

Following stalk section, two of the animals in Fig. 5.3 demonstrate maintained plasma TSH concentrations at the very upper end of the normal range (see Table 5.1) whilst fetal T_4 concentrations in these two fetuses fall well below the normal. It is difficult to find an explanation of such a change in thyroid sensitivity. However, these observations in stalk-sectioned animals must be borne in mind in relation to the changes described by these authors

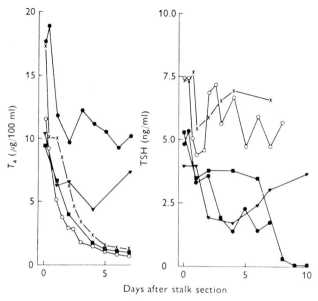

Fig. 5.3. Changes in fetal plasma TSH and T_4 following stalk section in the fetal lamb. Reproduced with permission from Thorburn and Hopkins [554].

immediately before the delivery of intact fetuses. In intact fetuses they observed a fall in fetal plasma T_4 which was not accompanied by any change in plasma TSH concentrations. One final point of note is that, following stalk section, fetal plasma TSH concentration is in the upper end of the normal range in 3 of the 5 fetuses. Liggins reports that pituitary stalk section did not lead to a decrease in fetal thyroid weight. The high fetal plasma TSH concentrations in some stalk-sectioned fetuses may be an index of leakage of TSH from cells within the infarcted area. A further possibility is that the TSH secreted after stalk section is a modified molecule in which biological and immunological activity do not bear the same relationship as in normal TSH. Alternatively, it is also possible that the technique of stalk section leaves an irritative focus which stimulates TSH secretion.

Considerations of fetal thyroid secretion of T_3 and T_4

The thyroid picture of elevated plasma TSH and T_4 concentrations with low circulating plasma T_3 concentrations in the chronically catheterised sheep fetus could be explained in several ways. Firstly, there may be decreased thyroidal T_3 secretion by the fetus. Alternatively, it may result from decreased peripheral monodeiodination of T_4 to T_3. In addition, increased clearance of T_3 from fetal plasma may be a factor. At the present stage some limited and preliminary evidence suggests that all three possibilities occur in the fetal sheep.

Administration of 50 μg TRH to fetal lambs of 125–135 days gestational age produces a marked increase in fetal TSH and T_4 (Fig. 5.4). Similar increments in plasma TSH and T_4 concentrations were observed after administration of

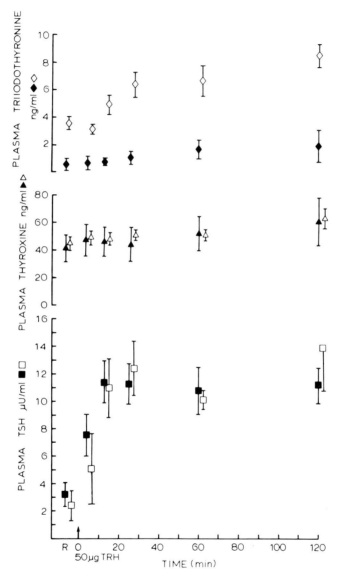

Fig. 5.4. Plasma TSH, T_4 and T_3 concentrations in 3 fetal lambs, 125–135 days gestational age (closed symbols) and 4, 5-day-old lambs (open symbols) after intravenous injection of 50 μg TRH. Thomas and Nathanielsz, unpublished observations.

the same dose of TRH to the 5-day-old newborn lamb. Fetal plasma T_3 rises in the first two hours after TRH but not to the same extent as in the newborn (see Fig. 5.4.). Since these observations were made in the two hours after TRH administration, the T_3 measured in the plasma would be of thyroidal origin rather than resulting from peripheral monodeiodination. Thus, this experiment shows that the sheep fetus can secrete T_3 but the $T_3 : T_4$ secretion ratio may be decreased. The concentrations of T_3 and T_4 in the thyroid of the fetal and adult sheep are similar [126]. The concentrations of iodothyronines in the thyroid gland of the newborn lamb remain to be estimated. The cause of the lower $T_3 : T_4$ secretion ratio demonstrated by the fetal thyroid when compared with the newborn makes an interesting region of speculation. It is known that in the adult, high $T_3 : T_4$ secretion can be achieved in humans with endemic goitre but it is uncertain to what extent this is due to iodine deficiency or to intense stimulation of the thyroid with TSH [370]. In the dog it has been shown that after TSH administration the amounts of both T_4 and T_3 released into the thyroid vein increase. However, the concentrations of T_3 were variable [555]. Chronic iodine deficiency leads to an increase in plasma $T_3 : T_4$ ratio but the endocrine picture is not the reverse of that seen in the fetus where the $T_3 : T_4$ ratio is low. TSH concentrations are elevated in both iodine deficiency and in the fetus [232]. There is a need for further work directly related to investigation of fetal thyroid secretion of T_3. In view of the dual origin of circulating plasma T_3 it will be necessary to catheterise the thyroid vein and observe the effect of various agents on fetal T_4 and T_3 secretion.

No studies have been made on the circulating concentrations of elemental iodine in the fetus. It is known that the placenta will concentrate iodide and therefore the fetal thyroid is exposed to a higher circulating concentration of iodide than the maternal gland [443]. From the evidence to be presented below it is clear that the unusual combination of T_3, T_4 and TSH concentrations seen in the sheep fetus cannot be wholly explained at the level of the thyroid gland.

It is worth noting that the PO_2 of thyroid arterial blood rises dramatically at birth. This change, together with pronounced hemodynamic changes and alterations in the autonomic nervous system, may play a role in the changes in the plasma T_3 and T_4 concentrations which occur in the minutes after birth. Newborn lamb plasma T_3 concentrations increase from less than 0.2 ng/ml at birth to as much as 3.0 ng/ml in 6 hours [445]. In view of the rapidity of the rise in plasma T_3 concentration in the newborn it is difficult to attribute the change to a sudden alteration in the rate of peripheral deiodination (see Fig. 6.1.). We have shown (Thomas and Nathanielsz, unpublished) that agents such as theophylline will increase the rate of appearance of T_3 in the plasma of thyroidectomised adult rabbits infused with T_4, but the time course of this increase is considerably slower than the increment in neonatal plasma T_3 concentration that follows delivery in the lamb. It should also be noted that the postnatal increment in plasma TSH is less than that achieved following the 50 μg TRH injection to the fetus shown in Fig. 5.4, and yet neonatal plasma T_3 concentrations rise by 3–4 ng/ml, whereas in the fetus the T_3 concentrations only increase by 1 ng/ml. These observations suggest that the fetal thyroid responds somewhat differently to TSH than the newborn thyroid.

Peripheral deiodination in the fetus
Erenberg and Fisher [reviewed in 193] have calculated the fractional T_4 to T_3 conversion rates of the thyroidectomised adult and fetal sheep using peripheral plasma studies after the intravenous injection of labelled T_4. They are 0.030 in the adult and 0.017 in the fetus. It should be noted that the clearance of T_3 from fetal plasma is faster than from adult plasma (5.5 h is the $t_{\frac{1}{2}}$ in the fetus whereas adult $t_{\frac{1}{2}}$ for T_3 is 7.0) [181]. This more rapid clearance will contribute to the lower fetal plasma concentration.

Studies of the conversion of T_4 to T_3 by different tissues in the fetus may yield more detailed information regarding differences in tissue deiodination. As mentioned earlier, monodeiodination to T_4 in the 5' position results in the production of T_3, whereas deiodination at the 5 position yields $3',5',3$-T_3 (reverse T_3 : rT_3), a potent antagonist of T_4. rT_3 has received considerable attention recently in the adult and the remarkable observations of rT_3 concentrations of 5 ng/ml in fetal plasma falling to 2 ng/ml or less in the first 3–4 days of neonatal life [126] may provide an explanation for the unusual combination of plasma TSH, T_4 and T_3 concentrations in the fetus. Since rT_3 is an antagonist of the action of T_4 in several bioassay systems [472], the presence of high concentrations of rT_3 in fetal plasma may explain the need for increased activity of TSH and T_4 in the fetus. If rT_3 interferes with the inhibitory effects of T_3 and T_4 at the pituitary and hypothalamic level, more TSH will be secreted, with the result that plasma T_4 concentrations rise to overcome the antagonistic effect of rT_3. More information is required regarding the binding of rT_3 to plasma proteins and hypothalamic and pituitary tissues. In fetal lamb plasma, free, non-protein-bound, rT_3 concentrations are half those of T_3 [126].

Many tissues have the ability to deiodinate T_4 in the 5 or 5' position. Little is known regarding the nature of these deiodinases or the factors responsible for their development and expression in the fetus. Grumbach and Kaplan [253] have suggested that the fetal pituitary is, in general, less constrained than the adult. In general terms, plasma growth hormone (GH), TSH, prolactin and ACTH concentrations are greater in the fetus than the adult sheep and the same difference probably exists between the adult primate and primate fetus. They explain these observations by suggesting that inhibitory inputs to the fetal hypothalamus are either not yet developed or are less important in the fetus. With regard to the fetal thyroid axis, an alternative, somewhat teleological, explanation is that the fetus has had to develop a peripheral mechanism to protect itself against the effects of excess circulating T_4 by producing more rT_3 than the adult. As we shall see in Chapter 8, in the case of GH there is some suggestion in the human fetus that, although the circulating fetal GH concentrations are elevated, as in the sheep, the somatomedin (SM) concentrations in human neonatal cord blood, as measured by bioassay, are less than those found in adults. Somatomedin is the term given to a group of compounds which are produced as a result of the action of GH. It is thought that in several instances the various forms of SM are intermediates in the actions of GH. This observation of high plasma GH and low SM concentrations suggests either end-organ resistance in the tissues responsible for the production of SM, if the absolute concentrations of SM are low (and not just its activity), or the possibility of some inhibitor of SM activity, if it should prove that the concentrations of SM are normal but its activity is low. Similarly for most of the latter part of fetal life, the adrenal cortex shows a relative insensitivity to ACTH (p.141). The fetal thyroid in contrast responds well to TSH by releasing T_4 and a peripheral inhibitor may be necessary.

The alternative explanation, mentioned above, would put the presence of excessive 5 deiodination in the fetus as the primary causative process. The resultant production of rT_3 might be expected to antagonise the action of T_4. If this antagonism by rT_3 also demonstrates itself by inhibiting the action of T_4 in exerting its negative feedback, the hypothalamo–hypophysial–thyroid axis will secrete more TSH and T_4.

Whatever may be the explanation of the fetal picture, it is clear that around the time of parturition there is a great change in the physiology of T_3 and rT_3. The increase in plasma T_3 after birth could be due to increased thyroidal T_3 secretion, increased peripheral 5' monodeiodination or decreased T_3 clearance. One possibility is that loss of the placenta at birth removes a tissue which is rapidly deiodinating T_3 to further degraded products or removing it from the circulation in some other way. Evidence for sulphation of T_4 by the placenta has been provided in the monkey [511]. However, the placenta does not appear to possess significant deiodination capacity [596]. In addition, placental permeability to T_3 is too low to be an appreciable source of T_3 loss before birth. It should be recalled that the ductus arteriosus and ductus venosus constrict rapidly after birth, resulting in increased pulmonary and hepatic blood flow. Increased deiodination in these vascular beds, particularly the liver, may play

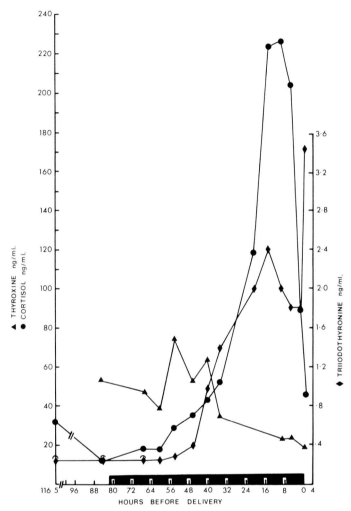

Fig. 5.5. Plasma cortisol, T_4 and T_3 concentrations in a chronically catheterised fetal sheep. Cortisol infusion (solid block) was begun at day 130 gestational age at a rate of 2.6 mg cortisol/day and thereafter doubled at intervals to produce a rapidly rising fetal plasma cortisol concentration. Thomas and Nathanielsz, unpublished observations.

a role in the increase in plasma T_3 concentrations at this time. One interesting observation is that neonatal cord plasma T_3 concentrations are much higher in the neonatal foal – the foal fetus does not possess a ductus venosus – than the neonatal calf or sheep (Rossdale and Nathanielsz, unpublished). Concentrations may be as high as 5 ng/ml. Investigation of fetal T_3 concentration in the pig would be of interest, since this species also lacks a ductus venosus.

There is much evidence that cortisol affects the peripheral actions of T_4 and it has been shown that T_4 deiodination is increased after cortisol administration to the thyroidectomised rat [438a]. The increase in fetal plasma cortisol concentrations before parturition may be involved in the changes in T_3 and rT_3 described, possibly through effects on the induction of specific deiodinating enzymes. In an attempt to dissociate the direct effects of cortisol and those which result from delivery induced by cortisol we have studied fetal plasma T_3 concentrations during cortisol-induced delivery. Cortisol infusions can depress fetal plasma T_4 and raise fetal plasma T_3 several hours before delivery occurs (Fig. 5.5).

Thyroid changes immediately before delivery
Thorburn and co-workers report a fall in fetal plasma T_4 concentrations in the last few days of intrauterine life and state that the fall occurs over the last ten days of intrauterine life [304, 554]. It is often extremely difficult to assess the exact commencement of a continued change in hormone levels when day to day variation occurs. The values in Fig. 5.6 represent the mean of six animals in all cases, except at 135 days (two animals). It would seem that there is no significant change in fetal plasma T_4 concentration from 120 days (90 ng/ml) to 145 days (96 ng/ml) and that the fall occurs between 145 and 150 days. In the absence of any standard errors it is difficult to assess the fall in these last five days. Fig. 5.7 illustrates this problem with data from one animal. The day to day variation in a single animal – not always apparent from meaned data – shows that there may be a tendency to consistently lower values for plasma T_4 but this is confined to the last five days in utero and on one of these days the concentration has risen again to the earlier levels. As seen above, infusion of cortisol into sheep fetuses at 130 days gestation at rates producing increases in plasma cortisol equivalent to those seen at delivery, will produce a fall in fetal plasma T_4 concentration (Fig. 5.5).

The evidence presented below in the discussion of thyroid function in the fetal calf is in keeping with these observations and leads to the conclusion that any fall in fetal plasma T_4 before delivery is secondary to the mechanisms of parturition and probably reflects the effects of cortisol on the peripheral metabolism of T_4. In the largest series of catheterised fetuses reported, Mellor and co-workers [412a] demonstrated that in 10 of their fetuses plasma T_4 did not fall before delivery and in 11 fetuses a decline in fetal plasma T_4 concentration occurred in the last 3 days or more of gestation [412a]. These observations confirm that a fall in fetal plasma T_4 concentration is not a necessary prerequisite for delivery to occur.

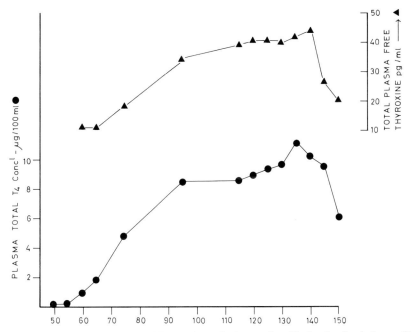

Fig. 5.6. Changes in plasma total T_4 and total plasma free T_4 in the fetal sheep. Data from Hopkins and Thorburn [554].

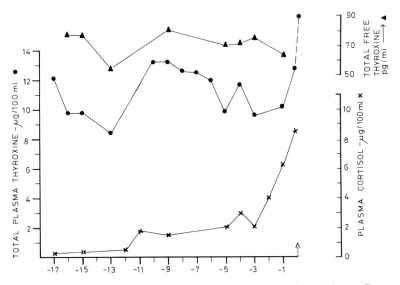

Fig. 5.7. Plasma total T_4, total free T_4 and cortisol in one fetal sheep. Reproduced with permission from Nathanielsz et al. [443].

5.3.2 Calf

Fetal plasma concentrations of T_4 remain steady in the chronically catheterised bovine fetus until three days before delivery [439, 444]. It should be noted from the outset that these values were obtained in 7 fetuses which delivered between 25 and 3 days prematurely, probably as a result of surgical intervention and the procedures associated with daily fetal sampling. The fall in fetal plasma T_4 which occurred over the last 3 days of fetal life was preceded by a fall in plasma TSH commencing about 24 hours previously. Plasma T_3 did not rise appreciably but the percentage of free T_4 in the fetal plasma did increase so that the total free T_4 in fetal plasma was maintained. These thyroid changes occur after the rise in fetal plasma cortisol has begun [439, 444].

Thyroidectomy of fetal lambs at 130 days or earlier does not prevent delivery. It would be useful to know if the normal rise in fetal cortisol that occurs at term also precedes delivery of the thyroidectomised fetus. However, taken together, the data from the fetal lamb and calf are in keeping with the conclusion that changes in fetal thyroid variables immediately prior to delivery follow the changes in the fetal adrenal axis in time rather than precede it. A temporal relationship does not prove a causal connection. However, since premature induction of parturition following the administration of ACTH or cortisol to the fetal calf, also produces similar changes in fetal plasma TSH and T_4 to those described during spontaneous calving, it would appear that the thyroid changes are secondary to the changes in plasma cortisol. In both the lamb and the calf it has been suggested that these observations reflect an inhibitory effect of glucocorticoids on the thyroid axis, probably by suppressing TSH secretion [439]. Fetal lamb plasma T_4 and TSH concentrations rise immediately after birth, at a time when plasma cortisol concentrations are also elevated. These observations suggest that the stress of parturition can directly overcome the inhibitory effect of the rising plasma cortisol concentrations.

CHAPTER 6

The fetal thyroid. II. The thyroid in the rat, human and sub-human primate fetus

6.1 Function of the thyroid axis in the fetal and neonatal rat

6.1.1 Development of thyroid function in the fetal rat

As mentioned in the previous chapter, the developing thyroid of the fetal rat goes through the same stages as in other mammals. T_4 has been measured in fetal plasma on day 18 [342]. In view of the latent period of action of T_4, even in species such as the rat, which have a very rapid metabolic rate, there may be great difficulty in demonstrating effects of T_4 deficiency in the rat fetus using the conventional procedures following thyroidectomy or goitrogen administration. At the present time there is no published data of T_4 and T_3 changes in plasma around the perinatal period in the rat. Our unpublished observations are that newborn plasma T_4 concentrations as well as T_3 concentrations are low (Fig. 6.1). If the fetal picture is similar to that shown in the newborn with low circulating plasma T_3, it would appear that the intrauterine phase of cerebral development in this species, as in the sheep, calf and human, can occur in the presence of lower T_3 concentrations in plasma than those in the adult animal.

With the availability of methods of assay requiring very small volumes of blood, investigations of changes in plasma TSH, T_4, T_3 and rT_3 in the rat during the perinatal period are now possible. They should enable us to relate thyroid activity to the distribution, uptake and metabolism of iodothyronines. It is clear that intracellular modification of hormones plays an important role in the actions of androgens in development, the same will probably prove to be true of the thyroid hormones. The fetal and neonatal rat should prove a useful model in which to investigate tissue uptake and metabolism of T_4 and T_3.

TSH and TRH have been shown to be present in the fetal rat pituitary and

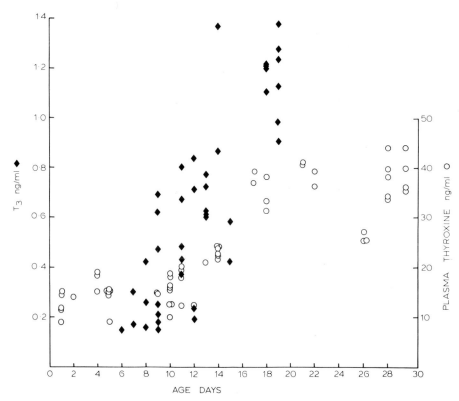

Fig. 6.1. Plasma T_3 and T_4 concentrations in the newborn rat from birth to day 30. Day 1 is the day of birth.

hypothalamus by day 19 of fetal life [141]. However, concentrations of TRH were probably about 20% of adult concentrations. Adult concentrations were only achieved by two weeks after birth. Fetal pituitary TSH concentration undergoes a 10-fold increase between days 18 and 22 [141, 155]. Following thiouracil drug administration in the intact fetal rat, various signs of increased secretion of fetal TSH can be noted, thus showing that the feedback mechanism for thyroid regulation develops in utero [313, 341]. The response to propylthiouracil can occur in the encephalectomised fetus, although the response is slightly diminished, suggesting that the pituitary is to some extent independent of the hypothalamus. TSH does not cross the placenta in the rat since fetal thyroid weight and iodide uptake are not affected by maternal thyroidectomy [273, 347, 368, 452].

Some evidence exists for release of TRH by the fetal rat hypothalamus, at least under certain conditions. The 24-hour uptake of iodide and circulating plasma T_4 concentrations are depressed to a small extent in the encephalectomised rat fetus, compared with the intact animal [341, 342]. Administration of 12 μg TRH to the intact rat fetus at 21 days gestation (approximately 1.2 mg/kg) will produce histological evidence of thyroid activation within 15 min [156]. This is a rather high dose; in the human the standard TRH stimulation test is based on the administration of only 200 μg TRH to the adult (approximately 3 μg/kg). In the same study, repeated TRH injection into the pregnant rat (500 μg given at 2-h intervals) produced an elevation of fetal plasma TSH and reduction of fetal pituitary TSH in the first two hours after delivery. These experiments suggest passage of TRH across the placenta. However, caution must be exercised, since placental characteristics are changed by anaesthesia and labour. Also, as discussed in relation to similar experiments in the fetal sheep preparation, it is very unlikely that TRH or other hypothalamic factors exist in high concentrations in the maternal peripheral plasma, and certainly never in the concentrations achieved by the injection of 500 μg two-hourly.

6.1.2 Placental transport of iodothyronines in the rat and guinea pig

In an attempt to assess maternal to fetal transfer of T_4, plasma T_4 concentrations in thyroidectomised fetal rats from normal and thyroidectomised mothers were compared [239, 240]. Fetal plasma total T_4 concentrations were higher in the thyroidectomised fetuses of normal mothers when compared with those of thyroidectomised mothers. If no allowance is made for any differences in thyroxine binding and utilisation rates, it would appear that the mother may contribute about 25% of the T_4 in the fetal pool. However, the results are open to an alternative explanation. The lower fetal T_4 concentrations in thyroidectomised pups of thyroidectomised mothers, when compared with intact mothers, may be caused by the loss of T_4 from the fetus to the thyroidectomised mother. The maternal thyroid pool is very large compared with the fetal pool and although the mean total T_4 was slightly higher in the thyroidectomised mothers than their thyroidectomised fetuses (0.7 to 0.3 ng/g plasma), there was no investigation of the transplacental gradient of free T_4. In these experiments it would only be necessary for there to be a slightly higher proportion of T_4 in fetal blood that is non-protein bound for there to be a positive feto-maternal gradient for total free T_4. If such a difference in total free T_4 should exist, T_4 would be lost into the large maternal pool and, as a result, the total T_4 in these thyroidectomised fetuses of thyroidectomised mothers would be less than in thyroidectomised fetuses from intact mothers. These authors give values of a butanol-extractable iodine (BEI) plasma concentration of 2.6 ng/g plasma in the fetal rat at 21 days gestation. Concentration in maternal plasma at this age was 6.2 ng/g plasma. BEI is a very close approximation to total T_4.

Experiments such as this have given rise to the general conclusion that the rat placenta is permeable to T_4. It is unfortunate that studies with radioactive isotopes have not been undertaken in this species. It is to be hoped that future studies will attempt to relate activity to circulating plasma concentrations of the hormones of the axis, including concentrations of free hormone.

In summary, by the end of gestation it appears that all the links in the thyroid axis are capable of action in the fetal rat [231]. It also appears that, as with the sheep fetus, the fetal rat thyroid is more active than the maternal thyroid [438]. Radioiodine uptake by the thyroid shows a marked peak immediately following delivery suggesting activation of the axis at birth as in the human, sheep and calf [220]. However, the postnatal patterns of changes in T_4 and T_3 are not similar to those in the sheep and calf. Further work needs to be done to clarify the similarities and differences between these species in order to assess the possible usefulness of the rat model in studying difficult aspects of fetal thyroid development.

Radioactively labelled T_4 and T_3 have been more extensively used in the guinea pig than the rat to study placental transport of thyroid hormones. Only studies in the maternal to fetal direction have been performed and harvesting the fetuses can only be undertaken under anaesthesia, which may well increase placental permeability. The general conclusion is that transfer of T_4 and T_3 from the maternal to fetal blood is limited [289, 398]. The administration of large doses of unlabelled T_4 and T_3 to the pregnant guinea pig has yielded conflicting information regarding their ability to cross the placenta and suppress the fetal thyroid axis. These studies were generally performed before simple methods of estimation of maternal and fetal T_3 and T_4 concentrations were available. In one study in which it was concluded that T_4 could cross the guinea pig placenta and inhibit propylthiouracil-induced thyroid hyperplasia, pregnant guinea pigs were administered 25 µg T_4 daily. Since the daily T_4 replacement dose for the thyroidectomised rat is 3 µg/100 g this dose is unlikely to have increased plasma T_4 concentrations above the physiological range [465]. Stable T_3 has been reported not to cross the placenta in more than limited amounts in this species [154, 475].

6.2 The human fetus and neonate

The developmental progression of thyroid differentiation, iodide accumulation, thyronine synthesis and secretion in the human fetus follows the usual sequence shown in other species. Much of the work clarifying the changes in fetal hormone concentrations has resulted from the laboratory of Dr. D.A. Fisher. TSH can be measured in the pituitaries of human abortuses removed at elective abortion as early as 14 weeks of gestation. An abrupt increase in both pituitary TSH concentration and content begins around 18 weeks [208]. Fig. 6.2 shows the

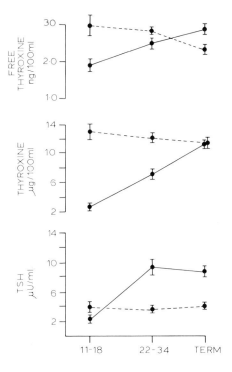

Fig. 6.2. Maternal (-----) and fetal (———) serum T_4, free T_4 and TSH concentrations from 11 weeks to term. Reproduced with permission from [208].

increase in fetal total serum T_4 which occurs between 11–18 weeks and term and the similar increase in free T_4. This increase is due predominantly to an increase in circulating fetal thyroxine-binding globulin (TBG) since free T_4 increases only 50%, whereas total T_4 increases about 500%. This increase in fetal TBG is thought to be due to increased circulating fetal estrogens. Fetal plasma T_3 concentration is very much lower than maternal T_3 concentrations (0.77 ng/ml compared with 1.93 ng/ml in one study) and rises in the latter weeks of pregnancy [209, 311]. This interesting observation remains unexplained but is of great importance in relation to the problems of production of T_3 in the fetus and the role of T_3 discussed in Chapter 5.

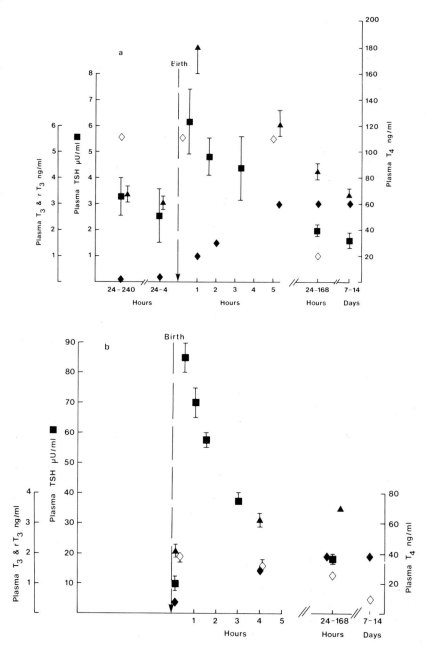

Fig. 6.3. Plasma T_3 ◆, rT_3 ◇, T_4 ▲ and TSH ■ concentrations in the fetal and neonatal lamb (a) and human (b). Data taken from and redrawn with permission from [125, 126, 181, 208, 210, 212, 213, 439, 443, 445].

In Fig. 6.3 these changes have been plotted with the various hormone measurements available from our work on the chronically catheterised fetal sheep. Measurements in the immediate neonatal period have also been included. In the minutes after birth there is an increase in neonatal plasma TSH (10-fold in the human and a doubling in the sheep), which is at least in part caused by exposure to cold, although other factors in the extrauterine environment are also of importance [211, 212]. It will be seen that there are considerable similarities between the human and the sheep in the perinatal period.

6.2.1 Placental permeability to iodothyronines

Whilst it is clear that the thyroid axis of the human and sheep act in a similar fashion in the perinatal period with regard to TSH and T_3, there is a divergence of view regarding placental permeability to iodothyronines in these two species. In the sheep, investigations in the second half of pregnancy by several groups of workers have shown that placental permeability to T_4 and T_3 is minimal [99, 134, 193]. Similar conclusions have in general been made in the human [104, 214, 255, 435]. However, some workers have concluded that T_3 crosses the human placenta more easily than T_4 [182, 482]. As mentioned below, there are considerable problems of experimental design in the human, as well as in the estimation of T_3 [207].

Investigations of placental permeability carried out around the time of delivery require very careful interpretation for two main reasons. Firstly, the integrity of the separating placenta is probably very different from that of the normal placenta in its undisturbed state. Secondly, if the end point of the experiment is the measurement of an endogenously secreted hormone (as T_4 in the example above), there are likely to be several other factors operating to change the concentration of that hormone in plasma. These factors may well mask any change caused by exogenous hormone administration. Although control groups can be found individual variation is high. In this situation it is probably preferable to use labelled hormone, though this expedient does not solve the problem of changes in placental permeability during delivery. In addition, it is very difficult to investigate the possibility of placental transport of iodothyronines in the very early stages of gestation before the fetal thyroid is capable of function. As discussed below, this may be a very critical period.

6.2.2 The role of iodothyronines in human cretinism

The human athyreotic cretin demonstrates deficiencies of mental and skeletal development as a result of thyroid deficiency in utero. Similar developmental problems occur even in fetuses with a thyroid in those areas of the world in which there is a chronic deficiency of iodine, particularly the mountainous areas of Peru, the Himalayas and New Guinea [316, 370, 476, 477, 478]. The classical myxedematous cretin demonstrates neurological abnormalities, including deaf mutism, impaired mental development, stunted growth and hypothyroidism. However, recent investigations of affected populations in these iodine-deficient populations living at high altitude, using very careful testing, have exposed deficiencies in individuals who are not apparently hypothyroid at the time of study. Also certain affected individuals demonstrate intellectual impairment without manifesting marked abnormalities of growth. It is tentatively concluded that these different clinical pictures represent the outcome of thyroid deficiency at different periods of development. It is known that brain development follows a definite pattern in the human, with sequential development of the various cellular elements at different fetal stages. A short period of thyroid deficiency may well affect certain neurological systems selectively, depending on the time at which the deficiency is present.

Related to this problem of critical phases of development in different systems is the nature of the thyroid factors responsible for the control of normal function. Various programmes have been carried out to treat pregnant mothers with iodine in iodine-deficient areas. From such studies it seems that the first twelve weeks of pregnancy are a critical period. The human fetal thyroid is not producing iodothyronines at this stage; indeed, even iodide trapping has not yet begun. If, as has been suggested, elemental iodine is of importance in the first few weeks of development, particularly in nervous tissue, the thyroid gland will not be competing for the iodide available in the fetal circulation. When considering the evolution of the thyroid gland it was suggested that a freshwater existence made an iodide-trapping mechanism necessary. No suggestions can yet be made regarding the role of iodide in cellular metabolism for which such a trap is necessary.

Thus, lack of elemental iodide or some non-thyronine compound may be responsible for developmental abnormalities which result from deficiencies dating from before the time at which the fetal thyroid can produce iodothyronines.

In summary, apparent thyroid deficiency can produce several different developmental abnormalities in the newborn human infant, ranging from simple deaf mutism to widespread deficiencies in neurological development and somatic growth. It is probable that the timing of exposure of the developing fetus to the thyroid deficiency will influence the exact nature of the developmental abnormality. Both iodide and iodothyronines may play important roles. Iodine is known to be concentrated by the placenta in several species and elevated fetal plasma iodide concentrations may be of importance in controlling development. The placental iodide trap is not dependent upon TSH. In addition there may be placental transport of iodothyronines in the early stages of pregnancy. The iodine-deficient pregnant woman may be unable to provide either iodide or iodothyronine in sufficient amounts to the fetus. Clearly an understanding of the pathogenesis of development in iodine-deficient areas will require more information on maternal thyroid function, placental transport of iodide and iodothyronines and fetal iodine and iodothyronine metabolism.

6.3 Experimental studies in the sub-human primate

Values for maternal and fetal T_4 plasma concentrations at 150 days gestation (0.88 of term) in the rhesus monkey are 96 and 72 ng/ml [359]. However, plasma binding of T_4 is less in the fetus than the mother, with the result that, as in the human and the sheep, the concentration of dialysable T_4 is greater in the fetus than the mother. One index of dialysable T_4, the 'free thyroxine index', gives values of 8.5 for the fetus and 7.7 for the mother at 150 days [359].

If, as appears likely, the fetus is exposed to a higher concentration of physiologically active T_4 in the later stages of gestation it is of interest to assess the degree of fetal autonomy in this species. Within the limitations imposed on placental transfer experiments in the monkey, the general conclusion would appear to be that transfer is limited. Administration of labelled T_4 to pregnant monkeys with subsequent measurement of fetal plasma radioactivity a day later, resulted in maternal to fetal concentration ratios of between 7 and 50 [467].

In these studies, the specific activity* of T_4 was measured on both sides of the placenta. In another study it was calculated that T_4 transfer from fetus to mother was greater than transfer from mother to fetus (9.2 ng/day compared with 0.4 ng/day. The placental transfer of T_4 from mother to fetus represented

---

* Specific activity denotes the proportion of the molecules which are in the labelled form and is derived from the formula:

$$\text{Specific activity} = \frac{\text{Quantity of radioactive molecules (curies)}}{\text{Total amount of the compound (grams)}}$$

It is a very useful index in endocrine studies, especially those involving constant infusion for the calculation of metabolic clearance rate and other endocrine features. In general the aim is to use radioactive preparations of high specific activity so that a large amount of radioactivity is added to the pool of the molecule under investigation whilst at the same time there is the addition of the minimal number of molecules of that compound to the total pool.

about 0.01% of the T_4 utilization by the fetus. Fetal utilization was calculated as 3.9 µg protein-bound iodine/day [369]. No weight was given for the fetus. If a fetal weight of 500 g is taken, T_4 utilization is 12 µg T_4/Kg/day or 230 µg T_4/ml/day. The corresponding figures for the sheep fetus are 40 µg T_4/Kg/day and 335 µg T_4/ml/day.

When radioactively labelled T_3 is administered to pregnant rhesus monkeys under pentobarbital general anaesthesia, there is minimal transfer of T_3 but iodide is freely transported [511]. Again, as with T_4, transport from fetus to to mother appeared to be greater than in the maternal to fetal direction. As the authors point out, the fetal injections were the equivalent of 85 µg T_3/Kg whilst maternal injections represented 13.6 µg T_3/Kg. Therefore the fetal injection would have had a greater effect on plasma binding of T_3 and this may be the explanation of the greater transport across the placenta when the T_3 was administered to the fetus. Following the administration of radioactively labelled T_3 to the fetus, an unknown labelled compound was found in fetal plasma. This compound could be hydrolysed by incubation with aryl sulphatase but not with β-glucuronidase. It therefore appears to be a sulphoconjugate of T_3. It did not appear in fetal or maternal plasma when labelled T_3 was given to the mother. It therefore seems that different pathways of metabolism of iodothyronines may exist in the feto-placental unit, compared with the adult. A similar compound is found in both fetal and maternal plasmas after T_3 administration to the fetal or maternal sheep [193]. Placental sulphoconjugation plays an active role in the metabolism of fetal estrogens. Further work is needed to elucidate its significance, if any, in fetal thyroid function.

Radioactive ablation of both the maternal and fetal thyroid at 71–88 days with the subsequent removal of the fetus for autopsy at 150 days has yielded useful experimental evidence regarding the role of the thyroid in the second half of gestation in the monkey [359]. It should be recalled at the outset that this type of model cannot be used for investigation of the important phases of development which occur about the time the fetal thyroid is beginning to trap iodide and elaborate iodothyronines. In these thyroidectomised fetuses both body weight and placental weight are decreased but not significantly. The fetal weight decreases less than the placental weight so that the fetal: placental weight ratio is increased. Fetal adrenal weight is decreased. This finding is of interest in the light of delayed parturition described in fetal lambs thyroidectomised at about the same gestational age [554]. Pituitary weight was increased. In this study, total brain weight was unchanged. This is a very crude index of neurological development and much more sophisticated studies have shown that fetal thyroidectomy causes several biochemical changes in the developing brain [298].

6.4 The role of the fetal thyroid in development

6.4.1 Growth

Growth may be defined as a permanent increase in size. This can result from either an increase in mean cell size and/or increases in the number of cells.

Surgical thyroidectomy of the fetal lamb at 0.55 of term results in a decrease in body weight of 33% of normal at delivery [305]. Similar experiments in animals with short gestation periods are difficult to interpret since it is impossible to allow adequate time to elapse after surgery for the circulating plasma T_4 ($\frac{1}{2}$-life 1–2 days) and tissue-bound T_4 to be cleared before delivery supervenes [268, 336]. Radioactive ablation of the thyroid in the fetal monkey at 0.5 of term also leads to growth deficiencies [298]. These findings are compatible with a rapid turnover of T_4 demonstrated in the sub-human primate [369]. It is clear, however, that the fetal thyroid axis is only one factor affecting growth and must be considered in relation to several other maternal and fetal factors [87]. The interested reader is directed to the review by Liggins entitled 'The Drive to Growth' [390].

6.4.2 Differentiation

In the multicellular mammalian organism every cell is initially derived from the fertilised zygote which has a fixed, specific and unique genetic complement. During development, the cells derived from this zygote develop the specific characteristics of the different tissues of the body. This process is called differentiation. At the present moment it is impossible to state whether the effects of thyroid hormones on growth and differentiation in the fetus are separate or linked. It will also be necessary to investigate the extent to which the effects of thyroid hormones are related to other hormones. Some indications of an interrelationship with cortisol at this period are given in Chapter 5.

The problem of investigation of the effects of thyroid ablation in short gestation species has been referred to. In addition, the effects of the use of fetal hypophysectomy or maternal administration of thiouracil drugs need to be interpreted with reservation. Hypophysectomy removes several important hormonal factors and thiouracil drugs will affect the peripheral deiodination of T_4 in the mother and the fetus as well as inhibiting thyroidal T_4 synthesis [195].

The established effects of thyroid hormones in the differentiation of four systems will be discussed: the type-II cell of the lung alveolus, bone, the skin, and the nervous system.

Differentiation of the type-II cell of the lung alveolus

T_4 (1 µg) injections into fetal rabbits at 24 days gestation accelerate the appearance of pulmonary surfactant in 2–3 days [595]. Investigations of the interaction of T_4 and cortisol on the differentiation of the newborn rat intestine suggest that the T_4 action may be secondary to its effect on glucocorticoid production. Administration of T_4 to newborn rats causes premature maturation of the small intestine but this is preceded by a T_4-stimulated increase in plasma corticosterone concentration [404].

The respiratory difficulty of fetal lambs thyroidectomised around 100 days gestation mentioned by Thorburn and Hopkins [554] may have been caused by a decrease in the production of pulmonary surfactant. No direct studies have been made on surfactant concentrations in fetal sheep which have undergone endocrine ablation or hormone treatment. We have thyroidectomised fetal lambs at 130 days and noted no respiratory problems at delivery at term. Human cretins do not have respiratory problems at birth and it is likely that any role T_4 plays in the regulation of surfactant production is not an indispensible one. Hypothyroidism of long duration probably depresses surfactant production and many other metabolic functions as a secondary result of more fundamental cellular processes.

Bone

Deprivation of T_4 in utero has clearly marked effects on bone development. In the hypothyroid human fetus, dysgenesis of ossification centres can be demonstrated as early as the seventh month of gestation [249]. Experimental studies on bone development in the perinatal period have shown that the decapitated rat fetuses exhibit retardation of bone growth [240]. Following decapitation at day 16 and assessment of bone development at day 21, it is possible to observe some bone changes similar to those seen as a result of more specific ablation of the thyroid in the fetal lamb. Because of the short gestational period and their immaturity at birth, many developmental phases are postnatal in the rabbit and rat. In the neonatal rat, thyro-parathyroidectomy leads to retardation of bone growth which is reversible with T_4 administration. It is therefore unlikely that the deficiency is due to lack of parathormone from the parathyroid glands [55].

Administration of goitrogens to pregnant sheep produces impairment of growth of the long bones of the fetus. It was initially proposed that these bone changes were secondary to impaired maternal health [383]. This was shown to be unlikely when a similar picture was obtained by Hopkins and Thorburn following fetal thyroidectomy. The technique used involved the removal of only one pair of parathyroid glands [305, 552]. Deficiency of ossification was most pronounced in the long bones, as was the appearance of regional lamellation of the bone shaft [383]. These changes were more prominent in fetuses thyroidectomised at 81 rather than at 92 days of gestation. These authors also made use of the twin preparation to demonstrate that the effect was not due to a secondary process in the mother subsequent to surgery, since the bone changes did not occur in one sham-operated fetus in a twin pregnancy in which thyroidectomy had been performed on the other twin. The basis of the failure of ossification in this situation is not understood. However, in relation to knowledge of the action of T_4 on protein synthesis and enzyme processes in development it is possible that T_4 is required for the synthesis of the correct mucopolysaccharide matrix upon which calcium can then be laid down.

The skin
In a comprehensive and detailed histological study, Chapman, Hopkins and Thorburn report the effect of removal of the fetal lamb thyroid at 80–100 days gestation [117]. The result is delayed development of all components of the skin. Administration of 200 μg T_4/day – a dose calculated to render the fetus hyperthyroid, speeds up the development of some of the components of the integument. These are predominantly the primary wool follicles, the sebaceous glands and epidermal keratinization. Following T_4 administration, little effect was demonstrated on the thickness of the wool fibre or on the development of the sweat glands and the dermis.

The nervous system
Hypothyroidism in the adult produces marked impairment of neurological function which is reversible with T_4 therapy. Recent investigations on iodine deficiency during human pregnancy have shown that there are different patterns of neurological deformity. For example, deaf-mutism may be the only major neurological deformity and occur in hypothyroid individuals with relatively normal somatic growth. Alternatively demonstrable intellectual impairment suggesting more widespread neurological damage may exist in the presence of normal somatic growth. A third picture is that of the classical myxedematous cretin who demonstrates neurological deformities, deaf-mutism, intellectual impairment and stunted somatic growth [317, 480]. An observation of therapeutic importance is that, contrary to earlier beliefs, some improvement may occur in older children after iodine therapy. However, the best results are obtained by early therapy and hypothyroidism acquired after the age of 2 results in little irreversible mental deficiency [530].

An understanding of these changes may be aided by the observation in the newborn rat that there is a critical phase of influence of thyroid hormones on neural development [185]. The degree of cerebral malfunction produced by neonatal thyroidectomy decreased with increasing age and if neonatal thyroidectomy is delayed until day 25 there is no detectable effect when the performance of the rat is tested at 120 days after being trained from 50 days in maze-learning tests. If replacement therapy with T_3 following thyroidectomy is begun by day 10, it is completely successful, but if it is delayed until day 25 there is no benefit to performance. A short period of hypothyroidism in the neonatal period results in irreversible alteration of the setting of the level of activity of the feedback system [34, 35]. Administration of excess thyroid hormone in the first two weeks of life will also produce resetting of the level of activity with a permanent reduction in pituitary and plasma TSH concentrations and a decreased TSH secretory reserve in response to thyroidectomy or TRH administration [65]. These changes in thyroid function are probably independent of the growth retardation which accompanies thyroid hormone administration in the newborn rat, since food deprivation adequate to produce the same degree of growth retardation does not produce a similar picture [32].

The experiments described above show clearly that the effects of thyroid (or iodine) deficiency are different if the deficiency is allowed to occur at different 'critical' phases of development. Human brain cell division occurs at two main periods, the first at 10–18 weeks affecting mainly the neuroblasts and the second, which is predominantly at five months postnatally, lasting well into the second year of neonatal life. The rat also has two major phases of brain development, one is late in utero, around days 18–20 and this can possibly be equated to the in utero phase in the human. A second, postnatal phase involving cellular proliferation, myelination and the development of a greater complexity of synaptic interrelationships, occurs before day 25 of neonatal life. These observations, whilst important, should be seen in the light of the probability that cell division and the quantity of cells in the brain is too gross an index. There are certainly more important qualitative and quantitative differences in terms of synaptic development. In the developing rat cerebellum it has been shown that T_4 deficiency leads to a slowing down of cellular development [450].

The detailed studies of Cheek and co-authors have demonstrated that T_4 deficiency in the developing monkey fetus produces a wide range of biochemical defects [298]. The major problem confronting investigators who wish to identify the primary effects of thyroid deficiency on neural development is the paucity of knowledge of the action of thyroid hormones at the cellular level. There are also well-documented interrelationships with other hormones known to be important in growth and development (see Chapters 5 and 8, dealing with cortisol and GH).

One study that highlights some of the problems has narrowed the critical phase to days 10–15 of postnatal life in the rat. The endpoint studied was the activity of brain succinate dehydrogenase (SDH). 75% of brain SDH is located in synaptosomal mitochondria. Thyroidectomy only affects brain SDH if it is performed before 10–14 days. Therapy begun on day 10 to rats thyroidectomised on day 1 restores brain SDH. If therapy is begun after day 15, restoration of brain SDH is incomplete but liver SDH can be restored by therapy at any time [568]. These observations suggest that the work of Eayrs [184–186] may eventually be explained on the basis that thyroid hormones (or iodine) are required for the progressive development and differentiation of enzyme systems vital to normal cerebral function. An effect of thyroid hormones has been shown on phospholipid synthesis, a vital component of membrane systems [346]. Further work should delineate the specific loci of action of thyroid hormones and the critical phases of development in man and the various experimental species under study.

CHAPTER 7

Structural and morphological development of the neurohypophysis

During embryological development, the fetal adenohypophysis, which is derived from Rathke's pouch, an upgrowth from the roof of the mouth, comes to lie beneath the hypothalamus and below the neurohypophysis which is a downgrowth from the hypothalamus. The fibres growing down the neurohypophysial stalk have their neurones in the hypothalamus. In the thyroid system, TRH is released from hypothalamic neurones and controls the pituitary secretion of TSH (see Chapter 5). In this chapter we are concerned with the function of two further groups of hypothalamic neurones, the supra-optic nuclei (SON) and the paraventricular nuclei (PVN).

The sequence of development of these structures is similar in all species studied to date. In the human, Rathke's pouch contacts the brain at 4 weeks gestation, the hypothalamic nuclei appear at 14–16 weeks, a solid neurohypophysis is formed by 16 weeks and active Gomori-positive stainable material is present in the hypothalamus by 19–20 weeks and in the neurohypophysis by 23 weeks. After 35–36 weeks there does not appear to be any further increase in the amount of stainable material [63]. Biological activity is present in the fetal sheep neurohypophysis before stainable material is present [12]. No increase in fetal neuropophysial oxytoxin can be demonstrated between 90 days gestation and term in the sheep whilst the pituitary arginine vasopressin (AVP) concentration doubles during this period. During this phase both adenohypophysis and neurohypophysis are growing steadily and therefore the total fetal stores of oxytocin and AVP are increasing in the second half of gestation. As discussed in general terms in Chapter 2, it is doubtful how much physiological significance we should put by observations of gland content. This is especially true with studies of the neurohypophysial hormones which are rapidly released in response to stresses of various different types. Recent advances in surgical and hormone assay techniques have permitted the use of the fetal sheep preparations for the accumulation of physiologically meaningful data regarding the development of the fetal neurohypophysis [119, 188, 225]. The observations show an acceptable degree of conformity between different groups of experimental workers. However, assay methods, particularly bioassay, require 1–2 ml of plasma. Frequent sampling may therefore constitute a haemorrhagic stress which may of itself result in secretion of octapeptides from the neurohypophysis, especially AVP. It is therefore a problem to interpret results obtained by too frequent samping. On the other hand the problem to be solved at the present time is that, since neurohypophysial hormones are released in 'spurts', infrequent sampling may miss important changes in secretion.

7.1 Chemical structure, synthesis and physiological properties

7.1.1 Oxytocin

The structure of oxytocin is shown in Fig. 7.1. Since most biochemists refer to the two cysteine residues which are joined by the S–S bond, as a single cystine amino acid, oxytocin is referred to as an octapeptide [183]. Destruction of this

```
Oxytocin              Cys—Tyr—Ileu—Gln—Asn—Cys—Pro—Leu—Gly—NH₂

Arginine vasopressin  Cys—Tyr—Phe—Gln—Asn—Cys—Pro—Arg—Gly—NH₂

Arginine vasotocin    Cys—Tyr—Ileu—Gln—Asn—Cys—Pro—Arg—Gly—NH₂
```

Fig. 7.1. The amino acid structures of oxytocin, AVP and AVT.

S–S bond results in loss of activity. The physiologically significant actions of oxytocin in the adult mammal appear to be confined to the female. It stimulates both the myoepithelial cells of the breast and the myometrial cells of the uterus to contract. Although various functions concerned with sperm transport have been suggested in the male, these have not been established. Oxytocin does have some antidiuretic activity but this is small when compared with AVP on a weight-for-weight basis. A possible role for oxytocin during parturition in both the male and female fetus is suggested below. The action of oxytocin on the myometrial cell is discussed in Chapter 12.

7.1.2 Arginine vasopressin (AVP)

The structure of AVP (Fig. 7.1) differs from oxytocin by only two amino acids. AVP affects salt and water transport in several mammalian tissues, especially the renal tubule. There is some evidence that the effects on sodium and water movements are separate actions in the frog skin and toad bladder which are the tissues in which water transport and ionic fluxes have been most extensively studied. The action of AVP appears to result in an increase in the activity of a membrane-bound adenylate cyclase in the target tissue with a resultant increase in intracellular cyclic AMP concentrations. There is also evidence that both alpha and beta adrenergic receptors influence the action of AVP on the renal tubule [526].

7.1.3 Arginine vasotocin (AVT)

Arginine vasotocin (Fig. 7.1) has been demonstrated in the fetal pituitary of various species [463, 539a]. In the fetal seal in which this molecule has been most extensively investigated, concentrations in the neural lobe are highest half-way through gestation. It is questionable whether AVT can be demonstrated in more mature fetuses. The evidence for the presence of AVT is based on the observation that the vasopressin fraction obtained after chromatography of pituitary extracts was more active when tested on frog bladder than could be accounted for by the measured AVP content. AVT is of considerable potential interest since its action results in the movement of water across both amniotic and allantoic membranes against an osmotic gradient [572]. Amniotic fluid volume is greatest around the mid-point of gestation in both the human and the sheep fetus. Amniotic fluid is hypotonic to both fetal and maternal plasma and AVT may play a role in the regulation of fetal water and electrolyte movements at this stage of pregnancy. Further work at various stages of gestation is, however, necessary, since certain observations are difficult to reconcile. Vizsolyi and Perks [571] demonstrated AVT in fetal sheep pituitaries between 81–91 days, whereas others could not demonstrate appreciable quantities of AVT between 91–107 days [13].

7.1.4 Pathways of destruction of octapeptides

Plasma removed from pregnant humans and monkeys possesses the ability to degrade both oxytocin and AVP. There are several enzyme systems (collectively referred to as oxytocinase) involved in this loss of activity. The major enzyme is cystine aminopeptidase (CAP). CAP is produced by the syncytiotrophoblastic cells of the placenta in primates but it has not been possible to demonstrate this activity in other species. Maternal plasma concentrations of CAP are low in early pregnancy in the human and rise throughout pregnancy with a similar time course to the changes in placental lactogen. Placental lactogen is produced in the same placental layer [54].

Tuppy [557] has reviewed the various enzyme systems involved in the degradation of octapeptides. Since oxytocin has a plasma half-life of 5–10 min and molecules secreted from the neurohypophysis take about 20 seconds to reach their target tissue, it is probable that plasma oxytocinase has little physiological significance. The presence of plasma and tissue systems which degrade these hormones is, however, a concern to anyone wishing to measure plasma hormone concentrations [244, 528]. Great care must be taken during sample collection to avoid loss of hormone activity. Too little attention is usually placed on the conditions under which blood is drawn, centrifuged and how the plasma is frozen and stored.

7.1.5 Neurophysins

Certain specific carrier proteins, neurophysins, have been extracted from the neurohypophysis. Neurophysin has been defined as 'cystine-rich protein, present in neurosecretory cells that specifically binds the neurohypophysial hormones' [303, 507].

Synthesised mainly in the hypothalamic nerve cell bodies, the octapeptides pass down the fibres of the supra-optico–hypophysial and paraventricular–hypophysial tracts attached to these carrier proteins. Neurophysin I binds oxytocin and neurophysin II, AVP. The binding is by non-covalent linkage. Problems with the assay of the octapeptides themselves have led to the establishment of assay systems for the measurement of neurophysin [384, 408]. Although the classical concepts of stimulation–secretion coupling for the release of hormones would suggest that the octapeptides and the neurophysins are released together, this may not be so. Great caution should be exercised in extrapolating from measured differences in neurophysin concentration to conclusions regarding the activity of the octapeptides. Even if the neurophysins and octapeptides are released together, there are certainly differences in the pools of distribution and rates of degradation of the different molecules. This has been demonstrated for the release of AVP and neurophysin in response to hypoxia (p. 101).

7.1.6 Methods of measurement

Both bioassay and radioimmunoassay data are available for oxytocin and AVP. AVP measurements in the fetus have generally been performed by radioimmunoassay although use has also been made of a bioassay based on the antidiuretic effect of AVP on the water-loaded rat, as measured by the resultant increase in urine conductivity. Oxytocin measurements have been made with radioimmunoassay and also the milk ejection activity bioassay using the 5–10 day postpartum lactating rat.

In interpreting the various results available in the literature it should be borne in mind that whilst the available radioimmunoassays are generally specific, there are considerable problems of plasma collection. As mentioned above, it is still uncertain whether plasma peptidases can significantly affect the results in maternal and fetal plasma of different species. This danger affects sample preparation to concentrate the hormone and remove non-specific effects. With respect to the bioassay systems, specificity is generally much less. For example, allowance should always be made for the milk-ejecting activity of any AVP present when oxytocin is being assayed.

7.2 Species to be considered

The species referred to in this chapter will be the fetal human, sub-human primates, sheep, rabbit and guinea pig.

7.3 Stimuli which release neurohypophysial hormones

Recent studies in the adult rat have demonstrated a dissociation in the release of AVP and oxytocin under different physiological situations. AVP and oxytocin are present in both the SON and PVN, thus disproving the older ideas of localization of oxytocic function in one nucleus and antidiuretic function in the other. However, haemorrhage and certain other forms of stress can be shown to stimulate the release of AVP, whereas suckling predominantly causes release of oxytocin [71, 264].

As mentioned above, there is at present a well-documented role for oxytocin only in the adult female. There are two separate neuroendocrine reflexes involved. The suckling reflex in which stimulation of the nipple produces a train of afferent impulses to the group of peptidergic neurones in the hypothalamus which release oxytocin. In the second reflex, the afferent input comes from the genital tract, particularly the vagina and cervix.

Release of AVP follows a much greater range of stimuli. The major physiological regulation is via afferent information from osmoreceptors located within the hypothalamic neurones themselves and from stretch receptors which are located on the arterial and venous sides of the circulation and are sensitive to changes in vascular volume. Our consideration of AVP in the fetus will require simultaneous investigation of the renin–angiotensin system in relation to response to haemorrhage. In addition, AVP is released in several stressful situations, such as hypoxia, even without a change in blood volume.

There is a striking similarity between the chemical structure of the octapeptides and the various hypothalamic releasing hormones whose structure is known (such as LH-RH which is a decapeptide). Since synthetic AVP will affect the release of ACTH from isolated pituitaries in vitro, it has been suggested that AVP may play a role in some instances when the adrenal axis is activated. The possibility that AVP is released in situations when ACTH secretion increases is discussed below (Chapter 9).

Recently it has been shown that oxytocin is released from the neurohypophysis in phases or spurts. Using lightly anaesthetised lactating rats, these spurts have been shown to be related to bursts of activity in hypothalamic neurones [574]. It is likely that neurohypophysial hormones are released in this phasic fashion at all stages of life and this fact should be borne in mind with respect to the reported concentrations measured in plasma. Similar phasic release of adenohypophysial hormones and the hypothalamic releasing hormones has been reported [376].

7.4 Oxytocin

Although no significant difference can be shown in the fetal pituitary content of oxytocin between 90 days and term in the sheep, there are significant changes in fetal oxytocin concentrations in plasma in the chronically catheterised sheep fetus. In 5 sheep fetuses, plasma oxytocin was less than 2 μU/ml in samples taken more than 50 hours before delivery, whereas in the two animals sampled in the last few hours before delivery of a live lamb, fetal plasma oxytocin concentrations up to 17 μU/ml were measured (Fig. 7.2) [224].

This increase in fetal plasma oxytocin concentration almost certainly reflects fetal secretion since, in the limited instances in which maternal and fetal oxytocin have been measured simultaneously, there is no correlation between the concentrations of the octapeptide on the two sides of the placenta in exteriorised fetuses [11]. Our unpublished data in chronically catheterised animals support independent fetal secretion during normal parturition, since there is no correlation between maternal and fetal plasma oxytocin concentrations. In addition, during the increase in fetal plasma oxytocin concentrations, which occurs when parturition is induced prematurely following the infusion of physiological amounts of cortisol into the fetus (Chapter 12), maternal and fetal oxytocin concentrations vary independently. Thus, the independence of fetal and maternal oxytocin concentration has been demonstrated in both acute and chronic sheep experiments.

In addition, the existence of high concentrations of oxytocin in cord blood in the newborn guinea pig [100a] and in human cord blood [120], in which there

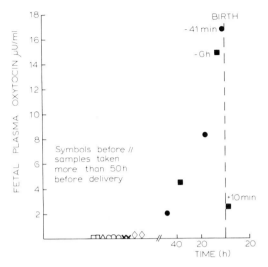

Fig. 7.2. Fetal plasma oxytocin concentrations in the last few days of intrauterine life. From Forsling, Jack and Nathanielsz [233].

is a positive umbilical arterio–venous difference, demonstrates that oxytocin is secreted by the fetal neurohypophysis during delivery. Cord blood concentrations are virtually unmeasurable in anencephalic fetuses. Human cord blood concentrations are also higher after normal vaginal delivery than in infants delivered by elective caesarian section when labour did not take place.

Radioimmunoassayable oxytocin can be measured in human amniotic fluid [513]. Concentrations are significantly higher in women in labour than in the amniotic fluid before labour. Of great interest is the demonstration by these authors of very high oxytocin concentrations in meconium-stained amniotic fluid. Using the same assay even greater concentrations of immunoreactive oxytocin were demonstrated in meconium itself. This is attributed to the high concentrations of oxytocin in amniotic fluid swallowed by the fetus. Oxytocin is also excreted in urine after birth. Whilst these are important observations in relation to the distribution of oxytocin and possibly other octapeptides within the fetal fluid compartments, it should be noted that no biological activity was demonstrated in these samples when they were tested on the isolated rat uterus. In addition, this data does not allow us to distinguish whether fetal oxytocin secretion precedes uterine contraction or is the result of stimulation of the fetus by labour.

The stimulus to fetal secretion of oxytocin during delivery may be pressure on the fetal head as a result of uterine contraction. The studies of Comline and Silver [138] have clearly shown that during normal labour in the sheep, fetal blood gas tensions do not fall until the very terminal phases of delivery; it is therefore unlikely in the sheep that hypoxia is a cause of oxytocin release by the fetus during delivery.

Maternal intravenous infusion of prostaglandin E_2 or $F_{2\alpha}$ for the induction of labour in the human, leads to release of oxytocin [243]. Prostaglandins are present in the fetal layers of the placenta. Fetal prostaglandins may play a role in uterine contraction (see Chapter 11) either directly or by stimulating the release of fetal oxytocin. This possibility should be investigated by the infusion of prostaglandins into the fetus. Fetal oxytocin may play a supportive role in parturition, since it has been demonstrated that oxytocin injected on the fetal side of the placenta is oxytocic. The administration of 100 mU of oxytocin to the fetal sheep will result in increased uterine activity (see Fig. 11.5) [442a]. Oxytocin administration to anencephalic human fetuses will also produce uterine contraction [300].

Further knowledge of the possible importance of fetal oxytocin in the mechanisms of parturition may be obtained from the study of the effects of fetal hypophysectomy on labour. Initiation of parturition has been accomplished by the infusion of ACTH into hypophysectomised fetal lambs [330]. However, one of the three hypophysectomised fetuses required delivery by caesarian section. The endocrine deficit following hypophysectomy is multiple and a possible effect of fetal neurohypophysial damage must be borne in mind. In addition, it has been observed in one study that dexamethasone, a synthetic glucocorticoid, will not induce delivery after fetal hypophysectomy [82]. As discussed in Chapters 9 and 12, these observations are best explained in terms of effects on the fetal adrenal. It is therefore possible that fetal secretion of oxytocin is impaired in hypophysectomised fetuses and the possibility that this deficit plays a role in the subsequent course of the delivery of hypophysectomised fetuses described above, must be borne in mind. However, it is this author's opinion with regard to presently available data that fetal oxytocin is not essential for the delivery of the fetus and any role it has is supportive. Proof of this opinion will require very careful monitoring of fetal oxytocin concentrations in different experimental models, especially the hypophysectomised fetus.

7.4.1 Placental transport of oxytocin

Since it can be shown that oxytocin administered to the fetal lamb is oxytocic, it is important to establish the route by which oxytocin gets to the myometrium. There are two possible experimental approaches to the investigation of transplacental passage of oxytocin. The first is to use radioactively labelled oxytocin. This method has not been used successfully to date. Although ^{14}C-labelled oxytocin is available it is in limited supply. ^{125}I or ^{131}I-labelled oxytocin is too unstable in vivo to permit experiments in which total radioactivity is measured after injection into either mother or fetus. If this technique is to be used it will be necessary to measure antibody-precipitable radioactivity.

The second approach is to inject unlabelled or stable oxytocin into mother or fetus with subsequent measurements of oxytocin in maternal and fetal circulations. In view of the rapidity of release of endogenous oxytocin as a reflex response it is imperative that recording of uterine contractions are taken during such studies. In addition, other forms of stress, which may lead to endogenous release of oxytocin, must be avoided. The various reflexes which stimulate the release of endogenous oxytocin have been discussed above. If the injection of exogenous oxytocin results in the contraction of the uterus, such contraction may reflexly stimulate the neurohypophysis and oxytocin may be released by mother or fetus. It is therefore probable that this method of approach cannot be used in late pregnancy when the myometrium is sensitive to oxytocin. This is a very real constraint on the investigation of this important problem, since it is very likely that placental permeability to oxytocin may alter in the last few hours of gestation when fetal oxytocin may play a role in uterine contractions. However, if oxytocin injected on one side of the placenta produces uterine contractions, any oxytocin subsequently measured on the other side of the placenta may have been endogenously released. Even the use of hypophysectomised fetuses and ewes would not totally remove the possibility of release from the hypothalamic neurones. In addition, a certain degree of care is required when investigating placental permeability in different experimental animal models. Degenerative changes have been demonstrated in the chorionic epithelium of the sheep placenta at term. These changes, which may be accompanied by alterations in placental permeability, occur after delivery in normal parturition [537] but commence prematurely in the cortisol-infused fetus [323].

There are several investigations which have produced data relevant to the problem of oxytocin transport across the placenta in the sheep, human and guinea pig [100a, 118, 223, 453]. Doubts must be exercised on the claim from one study that there is maternal to fetal passage of oxytocin in the sheep, since these experiments were carried out on exteriorised fetuses which had resting plasma oxytocin concentrations of between 1.2 and 3.0 mU/ml 20 days before term [453]. These high values were undoubtedly due to either fetal stress or methodological problems, since recent investigations show values for neurohypophysial hormones of 10 μU/ml or less at this stage in both acute and chronic preparations [7, 13, 224].

A study of transfer in the direction of fetus to ewe failed to show transfer between 114 and 134 days gestation [223]. However, this is not the critical period at which fetal oxytocin might be expected to stimulate the myometrium. In the guinea pig, samples have been obtained from the newborn pups within 2–3 min after delivery. Measurement of oxytocin concentration in these newborn guinea pigs revealed that the concentration increases with successive pups delivered in the litter. Whilst this observation is of great interest, it does not necessarily support the suggestion that 'the present observations showing increasing levels of oxytocin in fetal blood with successive deliveries, suggest that in the guinea pig, this oxytocin originates from the maternal circulation' [100a]. An equally acceptable explanation of high oxytocin concentrations in the later pups in the litter is that the later delivered pups have been exposed to a greater degree of the different stresses associated with delivery. Again the possibility of high values of plasma oxytocin concentration due to spurt release must be borne in mind.

It is clear that the role of neurohypophysial hormones in the fetus will be intensively investigated in the next few years. It is to be hoped that experiments involving injection of stable oxytocin will not only involve the monitoring of uterine activity but will also contain measurements of oxytocin on both sides of the placenta, not just the side to which transport is being investigated. If measurements of oxytocin disappearance are available on both sides of the placenta, together with fetal and maternal weights, some quantitation can be performed. In some studies in the literature, rough calculations suggest that more than 100% of the total dose has crossed the placenta. This observation, that more oxytocin is circulating than was initially injected, suggests endogenous release.

These questions of placental transfer of oxytocin have been considered at great length in an attempt to highlight the various problems such as spurt release, the effects of stress, positive feedback, influence of uterine contractions and further oxytocin release and changes in placental permeability with gestational age. It is also a further plea for precision in regard to the use of the word 'fetal' and an attempt to quantify fetal endocrinology.

At the present time it appears that there is no irrefutable evidence to counter the claim that at most stages of gestation the fetal and maternal compartments are independent with regard to circulating oxytocin. The route whereby oxytocin reaches the myometrial cell in late gestation remains to be elucidated.

Finally, in the adult animal it now appears that AVP and oxytocin secretion are controlled independently. The probability that the fetal neurohypophysial secretion of oxytocin is independent of AVP follows from a clear dissociation of changes in oxytocin and AVP in fetal sheep in response to haemorrhage and hypoxia [13].

7.5 Arginine vasopressin

In contrast to the lack of any change in the concentration of oxytocin in fetal sheep pituitaries between 90 and 143 days, pituitary AVP concentration doubles in this period [13]. In the chronically catheterised fetal lamb however, there is little change in circulating fetal plasma concentrations of AVP in the period from 20 days before delivery until about 3 days before delivery [7] (Fig. 7.3). More data is required before we can be certain of the exact time course of the pre-parturient increase in fetal AVP concentration. One of the problems is that too frequent sampling may itself cause a haemorrhagic stimulus to give rise to AVP secretion. Both acute and chronic investigations in the sheep suggest that the fetus is autonomous with respect to AVP secretion and that the placenta is impermeable to AVP [7, 526]. This is in keeping with the conclusions reached regarding placental permeability to oxytocin.

Kinetic studies performed in the chronically catheterised sheep clearly

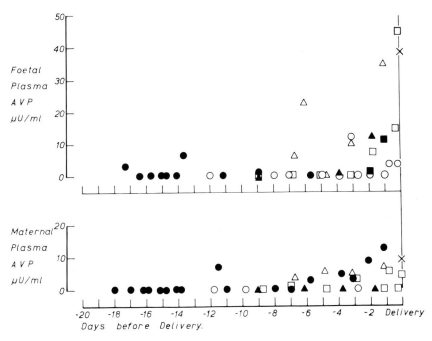

Fig. 7.3. Changes in maternal and fetal plasma AVP concentrations in the 20 days before delivery in the chronically catheterised fetal sheep preparation. Reproduced, with permission, from [7].

demonstrate that the overall blood production rate* for AVP is higher in the fetal lamb and the fetal rhesus monkey by a factor of about 100 when compared with the mother in each species [526]. The exact extent by which the fetal activity is greater than the mother is difficult to assess by the single injection technique used by the authors since, as they state, analysis of the maternal disappearance curve for AVP was very complex. In addition the mean resting fetal plasma stable AVP concentration of 49.1 µU/ml is somewhat higher than the concentrations found by other workers, generally less than 5 µU/ml [7, 504] and suggests an element of fetal stress. Furthermore, it is clear that oxytocin and AVP are secreted episodically, which leads to difficulties in the calculation of hormone blood production data. This problem is discussed in greater detail in relation to ACTH in Chapter 9.

Comparable data is not available in the human but high human cord blood concentrations of AVP have been demonstrated [119, 120, 307] and since the arterio–venous difference in the umbilical cord plasma is positive, it may be concluded that at least some of the AVP is not of maternal origin. As with cord blood oxytocin concentrations, cord blood AVP concentrations are lower in neonates delivered by elective caesarian section.

7.6 Physiological role of AVP in the fetus

AVP may possibly be involved in the following situations in the fetus: (1) responses to hypoxia, (2) responses to haemorrhage, (3) control of fetal body fluids, and (4) as a corticotropin-releasing factor.

The first three of these situations involve the reflex secretion of several hormones as well as certain cardiovascular responses. These hormone changes will be considered here in overall terms to permit the inclusion of data from endocrine systems other than AVP. This attempt is necessary to understand better the complex, multifactorial endocrine interrelationships in the fetus.

7.6.1 Release of AVP in response to hypoxia

When the pregnant ewe is made to breathe 9–10% oxygen the fetal lamb responds to the hypoxia produced with a rapid release of AVP [15, 504]. No difference is observed at gestational ages ranging from 110–140 days. The AVP concentrations in the fetus may rise to over 1000 µU/ml, whereas maternal concentrations only rise marginally. The increase in fetal AVP concentrations

---

* Calculation of blood production rates is discussed in Appendix I.

is less in fetuses in which the cervical vagosympathetic trunk has been cut [504]. In these experiments, radioimmunoassay for fetal and maternal neurophysin demonstrated release of neurophysin but these changes did not parallel the AVP concentration changes. The pathway by which hypoxia stimulates AVP release and the physiological significance of this response remain to be elucidated. Fetal ACTH and GH are also released by hypoxia and it appears that the renin–angiotensin system is inhibited. Fetal adrenalin and noradrenalin concentrations also increase during hypoxia [332]. The interrelationship of these hormones is considered in the next section.

7.6.2 Fetal endocrine responses to haemorrhage

'The place of the renin–angiotensin system before and after birth' has been excellently reviewed by Mott [427]. The picture which emerges from work in the fetal lamb and to a much lesser extent in the fetal monkey is that all the various components of this complicated system are present in the fetus. Renin, Angiotensin I (AI), and Angiotensin II (AII) have all been demonstrated in the fetus [88, 93, 105, 252, 290, 427, 543]. Plasma renin activity in fetal lamb plasma is higher than in the ewe whilst the concentrations of AII are similar in mother and fetus. This high plasma renin activity may reflect the low renal arterial pressure in the fetus or may represent an attempt by the fetus to make up for decreased conversion of AI to AII by pulmonary tissues [318]. Thus, more renin must be secreted to produce the same circulating concentrations of AII (Fig. 7.4). Similar disparities between GH and somatomedin, TSH, T_4

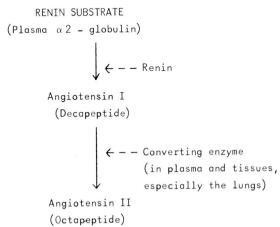

Fig. 7.4. The two major enzymatic steps in the conversion of renin to angiotensin II.

and T_3 are discussed elsewhere. Changes in the various components of the renin–angiotension system in the fetus occur in the absence of any maternal changes and it appears that in this system, as in other endocrine mechanisms, the fetus is autonomous.

Reduction of fetal blood volume by haemorrhage leads to renin release, even when blood volume is decreased by as little as 3% in the unanaesthetised fetal lamb [92]. Mott makes the following observation, 'In small animals, withdrawal of sufficient blood to permit assay of plasma renin can be a stimulus sufficient to increase plasma renin. Control experiments are therefore essential to exclude the effect of blood sampling per se in experimental procedures', [427]. The use of diuretics administered to the fetus in an attempt to decrease fetal blood volume is also a stimulus to the renin–angiotensin system.

As in many other systems, investigations of animals such as the rabbit, which are born in an immature state, have thrown useful light on endocrine development. The whole renin–angiotensin system is more active in the neonatal than the adult rabbit. This is in part due to a decreased inactivation of AII in various vascular beds in the neonatal rabbit. As a result, the neonatal rabbit is better able to withstand haemorrhage than the adult.

The role of the renin–angiotensin system in the fetus is not yet clearly defined. Plasma AII concentrations are low in lambs delivered by caesarian section when compared with plasma concentrations in lambs delivered vaginally. Increased renin activity in normal vaginal delivery may be due to decrease in blood volume as a result of haemorrhage at delivery. In addition, information is required regarding the effects of changes in the activity of the sympathetic nervous system during parturition in relation to the release of renin. In the adult mammal, sympathetic stimulation increases the release of renin. It has previously been considered that development of the sympathetic nervous system is not complete by the time of delivery even in species which are relatively mature at birth [137]. However, administration of isoprenaline, a β-adrenoceptor agonist to fetal lambs, produces tachycardia and hypotension [564]. This observation suggests a degree of tonic sympathetic activity in the fetus. In some respects the fetus appears to be more responsive to pharmacological administration of sympathetic agents. It is therefore likely that the fetal sympathetic nervous system and the renin–angiotensin system are capable of acting in concert to compensate for changes in blood volume and blood pressure in the fetus in utero, during delivery and in response to any haemorrhage at birth. Hypoxia has been referred to above as producing inhibition of renin release. During normal delivery of the fetal lamb, hypoxia is not a pronounced feature and will be more than offset by other stimuli which are stimulatory to the sympathetic and renin–angiotensin systems.

A full consideration of the short-term responses of the fetus to haemorrhage requires information on the sympathetic nervous system, as well as other cardiovascular and endocrine changes by which the fetus attempts to restore homeostasis. The cardiovascular compensatory mechanisms are well-developed in primates and other species born in a mature physiological state, such as the lamb. The endocrine mechanisms involve AVP, ACTH and adrenal catecholamines, as well as longer term responses of glucocorticoids and aldosterone in redistributing water and electrolytes.

There is no information on changes in secretion of catecholamines in the fetus following haemorrhage. This is due to current methodological difficulties in estimating adrenalin and noradrenalin without removing large volumes of fetal blood. It is clear that in newborn guinea pigs catecholamines play an important role in metabolic responses to cold [314]. Infusions of catecholamines and physiological stresses, such as fetal hypoxia, which might be expected to release endogenous catecholamines, result in increased lipolysis [312].

In acute experiments, AVP release in response to 7–20% loss of blood volume has been demonstrated in fetal sheep as young as 90 days gestational age. Maximal AVP concentrations were observed in fetal plasma following blood loss somewhat less than 40% [10, 14]. It is clear that AVP may play a role in any fetal response to haemorrhage but it is unlikely that fetal blood loss of this extent will occur often. It is important to note that in these experiments there was a dissociation of oxytocin and AVP response to haemorrhage. Oxytocin concentrations rose very little.

The original observations of Verney [570a] showed that AVP release was controlled by the response of the hypothalamus to the osmotic pressure of the blood circulating through it. There is some debate as to whether stimulation of AVP release occurs following hypovolemia in the adult sheep. The possible role of AVP in mineral and water control is discussed below. Haemorrhage results in a significant release of ACTH and GH in the fetal lamb as well as AVP. The possible role of ACTH is considered below. GH may stimulate metabolic responses which would be beneficial following haemorrhage. The increased secretion of GH poses the question as to the secretion of the other adenohypophysial acidophil, prolactin. In view of the possible role of prolactin in water and salt transport, it would be important to study prolactin release in response to haemorrhage.

7.6.3 Endocrine factors in the control of fetal body fluids

The sources of production of fetal body fluids and the mechanisms involved in water and solute fluxes have recently been reviewed by Barnes [38]. We must consider the possible actions of AVP, steroid hormones and prolactin at various epithelial surfaces. Fetal fluid composition depends on cellular activity at the placenta, chorion, skin, lungs, alimentary tract and within the fetal kidneys [86, 103] A carefully controlled experimental approach has only been conducted in the chronically catheterised fetal sheep in which allantoic, amniotic and urinary tract catheters can be placed in addition to fetal vascular catheters [38, 413, 414, 415].

Most of the early work using acute preparations has now been confirmed. In addition, enough data is available from stable chronic preparations to arrive at certain tentative suggestions regarding hormonal involvement. From the outset it should be noted that the human fetus has no fluid cavity comparable to the allantois of the sheep fetus. Fetal urine gains access to the allantoic sac via the urachus and to the amnion via the urethra in the sheep. Although fetal urine contributes to the fluid in both cavities, the composition of allantoic and amniotic fluids will differ as a result of the movements of ions and water at the other active epithelia mentioned above. Thus in the pig it has been demonstrated that Na^+ is actively pumped out of the allantois across the chorio-allantoic membrane [145]. Fig. 7.5 demonstrates the arrangement of fluid spaces in sheep.

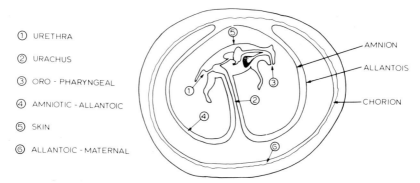

Fig. 7.5. Diagrammatic representation of the arrangement of the fetal membranes in the sheep. Reproduced with permission from Barnes [38].

During surgery for the implantation of fetal catheters, allantoic Na^+ concentrations fall and allantoic and amniotic K^+ concentration rises. These are the major changes observed in fluid composition, together with an increased amniotic fluid osmolarity. In the days immediately after surgery these changes are reversed [414, 416]. Urine osmolarity increases at the time of surgery and this increase is much greater in intact fetuses than in hypophysectomised fetuses [38].

The osmolarity change is probably caused by an increased secretion of AVP by the fetus. Contrary to earlier reports, AVP will act on the fetal and neonatal kidney to increase urine osmolarity. This effect is less, however, than that seen in the adult, since fetal glomerular filtration rate (GFR) is low compared with the adult – a feature which is probably due to the low renal blood flow in the fetus [38, 98, 251]. As a result there is a smaller filtrate available to be concentrated by the action of AVP. As term approaches there is an increase in urine osmolarity in the last few days of intrauterine life which correlates with the increased fetal AVP concentrations shown in Fig. 7.3. A circadian pattern of solute-free water clearance has been reported in the fetal sheep [98]. This may be due to a periodicity in the release of fetal AVP, or glucocorticoids.

Changes in fetal corticosteroids are however more likely to be the cause of any circadian variations. Glucocorticoids play a role in the control of GFR in adult mammals. Several other observations appear to link fetal glucocorticoids with changes in ion concentrations in the fetal body fluids. Mellor and Slater have demonstrated effects on fetal allantoic K^+ concentrations which are related to fetal blood glucose levels [416]. In addition, Mellor, Smith and Matheson [418] have shown that allantoic K^+ increases over the last few days of gestation in parallel with the increase in fetal glucocorticoids [46, 135] and it is likely that the K^+ changes associated with hypoglycaemia are caused by the increase in fetal glucocorticoids which accompanies hypoglycaemia [391]. A further observation which links the allantoic electrolyte changes described to fetal cortisol concentrations is the demonstration that premature induction of parturition by the infusion of ACTH into the fetus will raise allantoic K^+ [413].

As yet, no observations have been published of aldosterone concentrations in the chronically catheterised fetus. The role of this important hormone deserves much more attention in the fetus. AII can release aldosterone, as may high circulating concentrations of ACTH. It is possible that both of these systems operate immediately before delivery as well as during such physiological stresses as haemorrhage. Very high plasma concentrations (550 pg/ml) of aldosterone have been demonstrated in newborn guinea pigs [405a] and in human cord blood [52a]. Placental transport of aldosterone has been demonstrated [52a], but these observations are not without criticism, similar to those discussed in relation to placental transfer studies on oxytocin, since there are adequate stimuli at this time to cause endogenous release.

7.6.4 Is AVP a corticotropin-releasing factor?

Several groups of workers have demonstrated that AVP has actions on the hypothalamo–hypophysial–adrenocortical axis [401]. 'On the basis of the present and previous studies of the coupling between posterior pituitary hormones and ACTH release it appears that vasopressin has multiple effects on the adrenocortical system', states one group of eminent workers in this field [598]. They suggest that AVP may provoke release of endogenous corticotropin-releasing factor (CRF), may have innate CRF-like properties of its own, potentiate the response of the adenohypophysis to endogenous CRF and at high doses may even demonstrate ACTH-like properties.

The evidence on which these deductions are made was obtained in adult rats. In vitro studies with rat pituitary halves have shown that pituitaries incubated with 10–100 mU/ml AVP will release ACTH into the medium. This release can be inhibited by 0.05–5.0 μg/ml dexamethasone [216]. Whilst these observations are of great interest, it should be noted that the highest observed peripheral plasma AVP concentrations in the undisturbed sheep fetus are about 50 μU/ml (less than one percent of those used in these in vitro experiments). The possible relationship of pituitary portal and peripheral plasma AVP concentrations is considered below. In addition, the minimum dexamethasone concentrations used to inhibit the AVP effect are equivalent to about 25 times the circulating fetal cortisol concentrations. However, these are peripheral concentrations and no data are available for local concentrations of AVP within the adenohypophysis.

The failure of parturition to occur in fetal lambs in which a silicone rubber plate has been inserted between the hypothalamus and pituitary is generally explained on the basis of impairment of the passage of a signal from the hypothalamus which stimulates the release of ACTH from the adenohypophysis. If AVP is the humoral factor, pituitary stalk section would result in a failure of AVP to reach the adenohypophysis directly via portal vessels. In one interesting study in the monkey, individual pituitary portal vessels were cannulated under pentobarbitone anaesthesia. Portal blood AVP concentrations were about 14,000 pg/ml (4000 μU/ml) and peripheral plasma concentrations were 40 pg/ml (10 μU/ml) [601]. These concentrations are lower than those encountered in many experimental situations and further data are obviously required. In particular, the effect of administration of AVP to chronically catheterised sheep needs investigation. Infusion rates should be adjusted to keep peripheral concentrations at about 4000 μU/ml. It will also be necessary to demonstrate that such infusions produce the normal maternal and fetal endocrine changes which occur at parturition. An experimental approach of great value would be the implantation of cannulae for the localised infusion of AVP into the fetal pituitary, together with the estimation of adenohypophysial AVP concentrations at various stages of fetal life.

CHAPTER 8

Growth hormone, prolactin and placental lactogen

8.1 Hormonal similarities

Growth hormone (GH) and prolactin are related polypeptide hormones secreted by acidophil cells of the adenohypophysis. Various staining methods and specific immunofluorescent techniques have now quite clearly demonstrated that there are two different cell types each secreting only one of these two hormones [451]. These hormones exhibit species differences in their structure. Most of the clearly defined work on the physiology of these hormones in fetal development has been performed in the human, the sub-human primate, the sheep and the rat [493] and these are the species which will be discussed in this chapter.

The molecular weight of the monomer of GH in the various species studied is about 22,000 and there is some tendency for molecules of the hormone to associate. A detailed consideration of the structures of GH in different species is given by Wilhelmi [582]. Ovine prolactin has been extensively studied. It has a molecular weight of 24,000 and is related structurally to human GH and human placental lactogen (hPL). There are considerable functional similarities between these molecules in addition to their structural resemblance. hPL, produced by the syncytiotrophoblastic layer of the placenta [54] is indeed sometimes referred to as human chorionic somatomammotropin (hCS) thereby emphasising both its general growth promoting effects and its actions on the mammary gland. hPL has a molecular weight of about 20,000 and possesses considerable immunological and chemical similarity to human GH [227, 228]. Placental lactogens have now been isolated in several other species [355, 356, 407, 411]. hPL should not be confused with human chorionic gonadotropin (hCG). hCG is a glycoprotein of molecular weight 36,000, also produced in the syncytiotrophoblast. It is made up of a β-subunit which is specific to hCG and an α-subunit which is similar to that in other hormones (follicle-stimulating hormone (FSH), luteinising hormone (LH) and TSH). hCG is detectable in the blood at seven days after fertilisation. Estimation of hCG early in pregnancy is the basis of most pregnancy tests. hCG is active in maintaining the corpus luteum longer than would occur in the normal cycle converting it into the corpus luteum of pregnancy.

8.2 Growth hormone in the fetus

8.2.1 Sheep

Pituitary and plasma concentrations
GH has been identified in the pituitary gland of the fetal sheep as early as 50 days gestation by use of light microscopy [538] and 54 days using electron microscopical identification of specific cell granules [9]. Whereas plasma GH concentrations are 5–20 ng/ml in the pregnant sheep [575, 576], GH concentrations are greatly elevated in fetal plasma obtained under stable conditions from chronically catheterised fetuses. Fetal plasma GH increases from 40 ng/ml at 100 days to maximum values of 120 ng/ml at 140 days gestation [14, 49, 576]. A pronounced fall in fetal plasma GH concentration occurs in the few days before delivery (Fig. 8.1). The half-life of exogenous stable GH injected into fetal blood is 34 min. In lambs less than two days old it is 17 min and in 4–12-day-old lambs, 13 min [49]. However, when the clearance of GH is investigated using labelled GH, the metabolic clearance rate (MCR) is similar for fetal and newborn lambs*. Since these studies with labelled GH show that MCR is reasonably constant at different stages of development, blood production rates of GH are related to plasma GH concentration. However, the different results obtained in experiments with labelled and stable GH are worth remembering in relation to the various experimental systems used by different investigators.

The GH circulating in fetal blood is derived from the fetal adenohypophysis. This conclusion follows from the rapid fall to 1% of the normal concentration observed within three hours of fetal hypophysectomy (Fig. 8.2). In addition it can be demonstrated that labelled GH does not cross the placenta. The fetal pituitary may have a degree of autonomous function since following section of the pituitary stalk, fetal plasma GH concentrations drop to around 2–5 ng/ml (5% of the original level; Fig. 8.2). The exact extent of pituitary autonomy with

---

* See Appendix for the relationship between MCR and plasma half-life. The calculation of Blood Production Rate (BPR) is also discussed in Appendix.

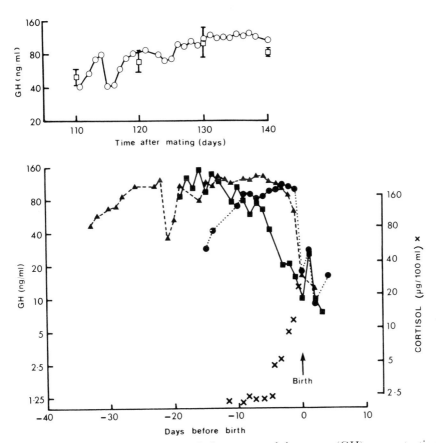

Fig. 8.1. Changes in maternal and fetal plasma growth hormone (GH) concentrations in the last days of gestation in the sheep. The increase in fetal plasma cortisol that precedes delivery is also shown. Reproduced with permission from Bassett et al. [49] and Nathanielsz et al. [422].

respect to GH is difficult to assess since pituitary stalk section is often accompanied by a degree of pituitary infarction (see p. 49).

Actions of GH in the fetal sheep
Hypophysectomy of the fetal sheep results in a definite, though not marked, retardation of growth whereas stalk section without any pituitary damage is followed by continued growth at roughly the normal rate [391]. In view of the multiple defects following hypophysectomy it is difficult to assess the extent to which the growth impairment reflects lack of GH. As discussed below, specific GH deficiency in human fetuses has little effect on growth.

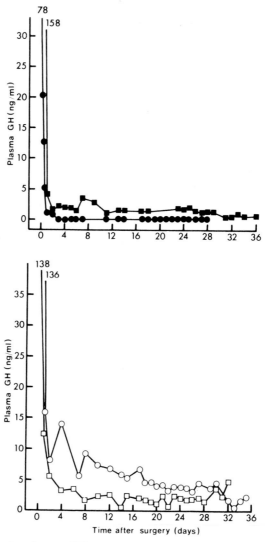

Fig. 8.2. Changes in plasma GH concentrations in the fetal sheep following fetal hypophysectomy (closed symbols) and pituitary stalk section (open symbols). Reproduced with permission from Wallace et al. [577].

Analysis of the effects of GH is complicated in three ways. Firstly, many of its actions may be mediated via intermediates. Factors affecting the generation and action of these intermediates may indirectly affect the action of GH. Secondly, the multiple and complex actions of GH on intermediary metabolism are inexorably interlocked with other hormones, especially insulin, glucagon, adrenalin, cortisol and T_4. Finally, the intermediary metabolism of the adult ruminant differs considerably from that of the non-ruminant. Investigations carried out in the fetal sheep should be interpreted in the light of observations in the adult sheep. Great caution should be applied in any extrapolation to the human fetus.

The fetal sheep demonstrates low circulating plasma glucose concentrations – approximately 15 mg/100 ml compared with maternal concentrations of 68 mg/100 ml and high plasma fructose concentrations of 70 mg/100 ml compared with unmeasurable concentrations of fructose in the maternal circulation.

In the adult mammal GH has effects on carbohydrate, protein and fat metabolism and its overall effect in any physiological situation will depend on the availability of its various substrates and the level of activity of insulin at that time [4]. Indeed, as discussed below and elsewhere [389], insulin is a potent growth-promoting factor. This was first pointed out by F.G. Young in 1940 [599]. Several interesting relationships have been demonstrated between insulin and GH in the fetus and the adult animal.

The kidney plays an important role in the clearance of GH from the plasma of the adult sheep [576], and the adipokinetic effect of GH is impaired in the absence of the kidney [575]. Plasma free fatty acid (FFA) concentrations are low in the fetal lamb as in the fetus in most species [138, 312]. These low FFA concentrations may represent an inefficient renal response to the high plasma GH concentrations. Whether this is disadvantageous to the fetal lamb in times of maternal food deprivation has not received experimental investigation. Battaglia and Meschia [51] consider that FFA do not play an important role in the metabolic balance sheet of the sheep fetus under normal conditions. However, they have demonstrated that infusions of noradrenalin at rates which double fetal blood pressure will elevate fetal plasma FFA concentration [326].

In many of its actions GH is considered not to be the final active agent but to act via intermediates [266]. One such intermediate has been named somatomedin (SM). It is now clear that there are a group of compounds, of polypeptide nature, which have been collectively referred to under this name [567]. Much of the difficulty arises from the wide variety of tissues and species that have been used in the various assays for SM activity. In general it appears that GH acts on various tissues, particularly the liver and kidney, to stimulate SM production. Preliminary investigations in the human suggest that plasma SM concentrations are low in the fetus. It has also been suggested that plasma SM concentrations feedback to inhibit GH secretion, thus constituting a type of negative feedback system. It is therefore possible that high fetal plasma GH concentrations reflect a lack of feedback inhibition by SM on GH release.

At the present moment the investigation of the role of intermediaries such as SM in the fetus is only just beginning and it is impossible to state whether renal and hepatic production of SM are low in the fetal lamb, or whether GH infusions in the intact, hypophysectomised or nephrectomised lamb will generate SM. SM production in the sheep has not received the same attention as in the rat. The chronically catheterised fetal sheep affords an ideal preparation for investigation of SM in the normal and surgically manipulated fetus. The interesting result of growth retardation in the nephrectomised lamb [552] is obviously the result of multiple endocrine and metabolic changes. The availability of high-titre specific anti-ovine GH antisera would permit the observation of the effect of specifically antagonising GH within the fetal circulation. It has been suggested in the human that GH is necessary for insulin response to hyperglycaemia [292]. This suggestion could be tested in the fetal lamb infused with anti-GH antiserum.

Control of GH secretion in fetal sheep
Five factors will be considered here: plasma glucose concentration; catecholamines and sympathetic nervous system; higher nervous activity; glucocorticoids, and lack of SM.

Fetal plasma glucose concentration Insulin-induced hypoglycaemia produces a release of GH in the adult sheep [575] as it does in the adult human [360]. Fetal plasma GH concentrations have not been tested in response to hypoglycaemia, nor is it known whether the very high fetal plasma GH concentrations represent maximal pituitary secretory capacity.

The plasma GH response to hyperglycaemia is variable postnatally [43, 400]. In 6-day-old lambs hyperglycaemia lowered plasma GH concentrations whereas in 28-day-old lambs GH concentrations were increased. Bassett and Madill [43] observed that infusions of glucose into the fetus, which led to plasma glucose concentrations of 40–80 mg/100 ml, lowered plasma GH concentrations in some fetuses although there was no significant negative correlation between plasma GH and glucose concentrations. One of the problems with the interpretation of these investigations is that the investigator has of necessity to measure plasma glucose concentrations. GH secretion probably reflects the availability of glucose within the cells of the hypothalamus responsible for monitoring the current physiological status with respect to their particular source of energy [282]. Measurements of plasma glucose concentration may not entirely reflect the activity of these cells.

Catecholamines and sympathetic nervous system Infusion of isoprenaline into fetal sheep will depress plasma GH concentrations [49]. Similar effects have been demonstrated in the adult sheep [49]. High plasma GH concentrations in the fetus may therefore reflect immaturity of an inhibitory influence, possibly of an adrenergic nature. Little is known regarding development of catecholamine and sympathetic nervous function around the time of delivery. The recent demonstration that infusion of 0.4 µg adrenalin/Kg/min to the sheep fetus produces similar circulating plasma concentrations as found in the sheep in response to hypoxia [332], is a useful guideline to investigators as to the rates of infusion which should not be exceeded. However, it must be clearly borne in mind that this is a measurement of plasma adrenalin concentration and may not represent postsynaptic release of catecholamine too closely.

Higher nervous activity As discussed in greater detail with relation to ACTH secretion, adenohypophysial hormone secretion in the adult is phasic. Bursts of activity are related to several external and internal stimuli which vary from species to species and also under different physiological conditions. One factor of importance is the sleep–wakefulness state of the individual. The majority of daily GH secretion in the adult human occurs during deep sleep associated with slow wave activity of the electroencephalograph (EEG). This slow wave EEG pattern usually occurs during the first two or three hours of the night only, with a subsidiary period of this type of activity shortly before waking. In the sheep fetus however, slow wave sleep occurs for about 60% of the day [163]. This is probably a considerably longer period than occurs in the adult sheep. This increase in the period of time spent in deep sleep may play a role in the production of high fetal plasma GH concentrations. It should be noted that the relation of GH secretion to EEG patterns has been worked out in the adult human and that even in the same species, the human newborn does not show the same EEG patterns as the adult. However, there are sufficient points of interest to prompt further attempts to relate adenohypophysial secretion to the level of nervous activity and development.

Glucocorticoids Although there is some debate, in several species glucocorticoids have been shown to inhibit GH secretion [488]. From 100–140 days gestation fetal plasma cortisol concentrations, both total and free, are very similar to adult concentrations. Just before parturition, the 10-fold increase in both moieties of cortisol (see Chapter 10) may be connected with the rapid fall in fetal lamb GH concentrations (see Fig. 8.1). GH discharge in response to hypoglycaemia but not to amino acids is inhibited by glucocorticoid [488]. In the adult sheep amino acids will stimulate the release of GH and prolactin. Different amino acids produce different secretion patterns of GH, prolactin and insulin [160]. No investigations of the effects of amino acid infusion have been carried out in the fetus.

As discussed in relation to the possibilities of T_4–cortisol interactions in the fetus (Chapter 5), it is important to establish the exact time relations of the rise in fetal cortisol and the fall in fetal GH concentrations before birth. Fig. 8.1 shows that the fall in GH definitely occurred in the last two or three days of gestation in two out of the three fetuses. It would therefore appear to follow the increase in fetal plasma cortisol concentration.

No evidence is available in the sheep as to whether the fall in GH before parturition is the consequence of the rise in cortisol, as is suggested for any changes which occur in the fetal thyroid axis, or whether they demonstrate an independent maturation. Observation of plasma GH concentrations during cortisol-induced delivery at 120–130 days would assist in differentiating between these possibilities.

8.2.2 Human

Despite the problems regarding the nature of the material available for study, a very clearly defined picture of the physiology of GH in human fetal development has emerged from the work of Grumbach and Kaplan [253]. We should however bear in mind the statement of these authors, 'We recognise the limitations in the measurement of the hormone content of an endocrine gland and not its flux, in the interpretation of the serum concentration of a hormone without knowledge of possible differences between the foetus and child in metabolic clearance rate and in the assessment of foetal age, and the potential or unrecognised artifacts related to the state of the foetuses studied'.

Pituitary and plasma GH concentrations
Fig 8.3 taken from Grumbach and Kaplan [253] is a summary of data from several sources. Data is available for activity at various levels of this endocrine axis. The general principles and some of the methods used to investigate the development of hypothalamic pituitary interrelation are discussed elsewhere (Chapter 4). At the outset it should be noted that GH does not cross the human placenta [381]. This conclusion has been drawn from experiments involving the administration of labelled and non-labelled GH to pregnant women immediately before delivery. The absence of transplacental transfer can be shown by the analysis of cord plasma to show that GH has not passed from mother to fetus.

Fetal pituitary content of GH is high, 1000 times that of prolactin, FSH and LH and the content rises steadily through gestation. This reflects the preponderance of somatotrophs in the fetal pituitary [190]. GH concentration in cord

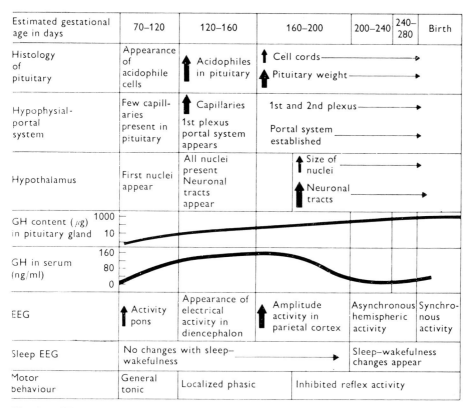

Fig. 8.3. The ontogeny of growth hormone secretion by the human fetus at different levels of the axis. Data reproduced with permission from Grumbach and Kaplan [253].

blood taken after delivery from the uterus is maximal at 160–200 days, after which time there is a profound fall in concentration. There is a noteworthy similarity between the very high absolute GH concentrations from human material obtained at about 160 days gestation (160 ng/ml) and the plasma GH concentrations in undisturbed chronically catheterised fetal sheep.

Grumbach and Kaplan propose a basic explanation to relate the known features of development of cerebral activity in the higher cortical areas, the appearance of various features of hypothalamic neurosecretion and the establishment of a functional pituitary portal system in the human fetus. They conclude that falling GH concentrations in cord blood at later stages of development reflect increasing inhibitory effects from the developing brain. The effect of adrenergic compounds on GH concentrations in the sheep fetus should be recalled. The possible inhibitory effect of glucocorticoids has been discussed in relation to the pre-partum fall in GH in the sheep fetus*. As discussed in Chapter 10, spontaneous vaginal delivery of the human fetus is probably also preceded by an increase in fetal cortisol secretion. Measurement of amniotic fluid cortisol concentrations in the human suggest that the rate of increase in cortisol secretion is much slower in the human fetus than the sheep fetus. As with the sheep fetus, it remains to be seen whether the fall in cord blood GH concentrations over the last few weeks of gestation is a consequence, either direct or indirect, of the increased fetal plasma cortisol concentration at this time.

GH has been detected in a cord blood of live-born anencephalics [253, 256]. In addition GH concentrations, although low at birth, may rise in the period during which the anencephalic survives [256]. In some cases a response to insulin hypoglycemia and arginine infusion is present although blunted. The implications of these findings to the endocrine status of the human anencephalic baby should be noted (see Chapter 12). It is quite clear that human anencephalic fetuses comprise a heterogenous group in terms of their endocrine capacity.

Actions of GH on fetal tissues

Laron dwarfs exhibit a familial defect in which immunoreactive GH concentrations are elevated but there is no increase in SM after exogenous administration of GH. These dwarfs are of normal body weight at birth and only slightly diminished body length. This observation increases the likelihood expressed above, that GH does not play an indispensible role in normal fetal growth [380].

Although no data can be provided at the present time regarding the possible significance of SM in the sheep fetus, SM has been measured in human pregnancy. The data available is of a very preliminary nature but these results give rise to some interesting speculations. Most SM bioassays** record the normal adult value as 1 [567]. In the pregnant human SM concentrations in maternal serum fall to about two thirds of the normal (0.67) and decrease even further at delivery to 0.26, which is in the hypopituitary range. One possible explanation is that other factors having the same metabolic actions as GH (possibly hPL, see below) are present in the maternal circulation. If this is so their resulting actions may feed back to inhibit maternal GH secretion. Since SM is measured by bioassay, these other factors obviously do not possess SM activity in the biological test system used, usually newborn pig costal cartilage.

---

* Somatostatin, an inhibitor of GH and insulin release has been identified in the pancreas of the human fetus from 12 weeks gestation [176].

** There are several different assay systems for the various somatomedins. For example, one which measures sulphation factor activity uses the ability of plasma to stimulate the uptake of radioactive sulphate into newborn piglet costal cartilage.

Cord blood SM concentrations are also low (0.54) despite elevated GH concentrations of 30.7 ng/ml [547]. The low generation of SM in the presence of high GH concentrations may be the result of renal or hepatic immaturity in the fetus. Some authorities have suggested that SM, as well as being the intermediate in GH's metabolic action, is itself the factor responsible for the feedback inhibition on GH secretion. If this is so, similar considerations might apply to GH and SM as discussed in relation to T_4 to T_3 interconversion and the overall level of activity of the thyroid axis (see Chapter 5).

One possible in vivo inhibitor of SM generation is estrogen. Estrogen does not affect the bioassay of SM directly when added to cartilage in vitro. However, estrogen therapy for 7–18 days to acromegalic patients and hypopituitary patients on GH replacement therapy diminishes plasma sulphation factor activity without altering plasma GH concentrations [578]. One further possibility is that there may be different tissue responsiveness to SM in the fetus. The problem of changes in tissue responsiveness during fetal life has received little attention in most endocrine systems. However, it has been shown that 16–20 day fetal rat cartilage is very unresponsive to sulphation factor. In this species maximum response occurs in the weanling rat with a subsequent decrease thereafter [275].

8.2.3 Some experimental observations in the sub-human primate

Chez, Hutchinson, Salazar and Mintz produced fetal hypophysectomy in the rhesus monkey *(Macaca mulatta)* by implanting Yttrium-90 in the fetal sella turcica at 123 days [122]. This procedure results in prolongation of pregnancy in those animals in which the fetus survived the operation by more than 10 days. These fetuses continued to grow, confirming the ability of the fetus to grow (though not necessarily to its full potential) in the absence of fetal pituitary hormones including GH. The two hypophysectomised fetuses which were delivered live had no histological evidence of residual pituitary tissue. Plasma GH concentration was unmeasurable and no increase in plasma GH was obtained in response to insulin. These findings should be compared to those reported above in the human anencephalic.

The same group of workers have reported that fetal GH concentration is 25 ng/ml at 140–149 days gestational age in blood taken from the interplacental vessel under Halothane anaesthesia. In the newborn monkey, plasma GH concentration is 7 ng/ml. Increasing fetal plasma glucose concentration by as much as 150 mg% for 2 hours at 142–148 days gestation did not suppress fetal plasma GH [422].

Hypoglycemia, or arginine infusions did not cause an increment in fetal plasma GH although hypoglycemia did produce a rise in GH in the neonatal monkey. These are the only data available in the sub-human primate. Although there was a lack of any significant change in fetal GH concentrations during the various experimental regimes it is of interest to note that the authors remark that there was an increase in fetal GH towards the end of the experiment in nearly all animals regardless of the protocol. Labelled GH did not cross the placenta in either direction.

8.3 Prolactin

8.3.1 Available measurements for hormone concentrations in the fetal pituitary and fetal body fluids

Human studies
It is only recently that any data has become available for prolactin concentrations in fetal tissues and body fluids. The content of prolactin in the human fetal pituitary is one thousandth that of GH and it increases continuously until day 200 of gestation remaining constant thereafter. In pituitaries from 24 human fetuses and sera from nine, prolactin was demonstrated by 90 days gestation in the pituitary and showed a significant relationship to GH concentrations [253]. At term, fetal serum prolactin concentrations were similar to maternal concentrations. High amniotic concentrations of prolactin up to 7000 ng/ml may be present in the first half of pregnancy. Concentrations in newborn cord blood are 250–350 ng/ml [166, 253, 560].

Sheep studies
Several reports have now appeared regarding the circulating concentrations of prolactin in fetal sheep plasma. In chronically catheterised sheep fetuses there is some variation between animals. From around 130 days gestation we have observed fetal concentrations of 20–60 ng/ml. Similar fetal plasma prolactin concentrations have been reported by other groups of workers. As with the chronically catheterised fetus, considerable individual variations appears to exist between animals when sampled under anaesthesia. In two sets of twins removed at operation under pentobarbitone anaesthesia, values of 54 and 12, and 70 and 5 ng/ml were measured at 122 and 138 days gestational age, respectively, [424]. At less than 110 days prolactin was below 3 ng/ml plasma, except in one fetus at 72 days in which the concentration was 3 ng/ml. In another study, fetal plasma prolactin was only measurable after 136 days gestation and concentrations ranged between 14 and 45 ng/ml [8]. The only study of prolactin concentrations in fetal sheep amniotic fluid is that of Wilson and co-workers

[583]. Concentrations about 10 ng/ml have been observed between 120 to 145 days gestation. Values are not available for the early stages of gestation corresponding to the very high concentrations in human gestation noted above*.

The administration of 50 μg TRH intravascularly to the fetus will elevate fetal prolactin concentration to a slightly greater extent than the increase obtained in plasma TSH concentration (see Fig. 4.3). This simple test may well prove of use as an index of pituitary function in the fetal lamb. Radioimmunoassay of prolactin is very simple and produces results within 2 days, thus enabling assessment of pituitary responsiveness as well as the basal level of activity from resting plasma concentrations.

Administration of TRH to the fetus does not result in an elevation of maternal plasma prolactin concentration. Similarly 200 αg TRH injected intravenously into pregnant ewes elevates maternal plasma prolactin concentration from 100 to over 400 ng/ml, whilst having no effect on fetal plasma prolactin concentration. It may therefore be concluded that prolactin does not cross the placenta The resting maternal to fetal concentration gradient is about 6 : 1, in contrast to slightly less than 1 : 1 for TSH and a minimum of 1 : 7 for GH.

From the experimental point of view, investigation of prolactin physiology in the fetus may throw useful light on the general features of development of fetal neuroendocrine systems. In the adult mammal the controlling hypothalamic influence on prolactin is an inhibitory one. It has been shown that bilateral lesions placed in the median eminence of female rats result in a tenfold increment in serum prolactin within 30 minutes. This elevation of prolactin continues for at least 5 months [580]. If a similar inhibitory control on prolactin exists in the fetus, stalk section may be followed by an increase in plasma prolactin concentration. Any increment in circulating prolactin following stalk section could play an important role in maintaining growth at a near normal level. As discussed elsewhere, it is difficult to assess the significance of the fall in GH following stalk section when accompanied by pituitary infarction (p.108). Unfortunately GH was not estimated in the stalk-section experiments of Liggins et al. [391]. Prolactin has been shown to have many of the actions of GH [451]. It would be of interest to know the effect of stalk section unaccompanied by pituitary infarction on fetal plasma prolactin concentrations as well as to study the actions of prolactin infusions in the hypophysectomised fetus.

---

* Too much emphasis should not be placed on the plasma concentrations of prolactin measured in the fetus. It has been demonstrated in the newborn lamb and ewe that blood production rate for prolactin is as much affected by MCR as by plasma concentration [161].

Important osmoregulatory effects of prolactin have been demonstrated in fish [451]. In the adult sheep, prolactin decreases both water and sodium excretion [308, 309]. Whether this is a direct effect or due to an interaction with AVP and glucocorticoids is unknown at the present time.

Prolactin has effects on the metabolism of cholesterol esters and progesterone precursors in steroid-producing tissues which have been reviewed by Nicoll [451]. He concludes 'A role of prolactin of major significance in vertebrate "function and organization" may be that of modifying the responsiveness of various organs to the trophic influence of other hormones as suggested previously by Nicoll and Bryant. Such modulation is evident for example in steroidogenic glands, male sex accessory organs and the mammary gland. Thus, prolactin could be viewed as a broad spectrum conditioner of the hormonal sensitivity of a wide variety of vertebrate organs'. This point will be reconsidered when discussing the function of the fetal adrenal cortex. Very little consideration has been given to the role of prolactin in the undisturbed fetal sheep preparation or in the hypophysectomised sheep fetus. Fetal hypophysectomy will prevent the fetal urinary electrolyte changes that occur at 135–148 days in the undisturbed fetus. Parturition would normally occur in the undisturbed fetus at this time [38]. It may be that the change in level of activity of the adrenal cortex is responsible for the effect of hypophysectomy on fetal fluid balance. However, the hypophysectomised fetus lacks prolactin and the role of prolactin on the formation of fetal fluids remains to be investigated.

Prolactin and response to 'stress' The studies of Lamming et al. [376] have shown that prolactin is released in an episodic fashion in the adult sheep and that secretion is very responsive to a variety of 'stresses'. Indeed, in the adult sheep, prolactin secretion increases in response to stress whereas GH does not [160]. It is difficult to envisage the immediate advantage to the adult animal of an increase in circulating plasma concentrations of prolactin. If prolactin does interact with the adrenal hormones, either cortical or medullary, then increased prolactin secretion in times of stress could be physiologically significant. The extent of any fetal prolactin secretion in response to different stresses remains to be elucidated. One of the major problems with the investigation of fetal responses to stress is to decide which 'provocative tests' of fetal adenohypophysial function are physiologically meaningful. For example, fetal inferior vena caval blood gases and pH do not change until the last minutes of intrauterine life [138]. As discussed on page 26 this does not necessarily preclude local changes in cerebral gas tensions. However, the uncertainty remains that whilst hypoxic stress to the fetus may occur under pathological conditions, other, as yet undefined, stresses may be more physiological.

It is quite clear that prolactin physiology in the fetus requires investigation on the basis of the remarkable concentrations in amniotic fluid, if for no other reason. However, the recent demonstration of the multitude of actions performed by this ubiquitous hormone should kindle great interest in its function during fetal life [451].

8.4 Placental lactogen

8.4.1 Human studies

Placental lactogen is a product of the fetal syncytiotrophoblast. Circulating concentrations in maternal plasma increase throughout gestation and have been used as an index of fetal well-being [550]. Low maternal hPL concentrations are associated with failure of fetal growth [144, 550]. Since it has been reported that placental lactogen is absent or low in term cord plasma, there has been a tendency to discount it from any role within the fetus [350].

Kaplan et al. [349] were able to detect hPL in all of 48 fetal sera obtained from hysterectomy. Concentrations were 2–240 ng/ml. Although there was no correlation with age, the levels were highest at mid-gestation. Term cord blood concentrations are only around 13 ng/ml. hPL concentrations in amniotic fluid are high – up to 930 ng/ml – and there is very little passage of radioactively labelled hPL across the placenta, which is in keeping with the observations on prolactin and hGH [229, 254, 350]. However, in relation to passage between fetal and maternal tissues, there are important considerations regarding the site of synthesis when comparing these three hormones. These are discussed below.

8.4.2 Sheep studies

The original idea that placental lactogens were only present in primate species is now known to be incorrect. Placental lactogens have been demonstrated in several species with various types of placenta. The pattern of changes in PL in maternal plasma are very different in the sheep and human (Fig. 8.4). In the sheep PL reaches a peak of 2000 ng/ml around day 110 and thereafter slowly declines until delivery [355]. Similar concentrations of PL were measured in the maternal and fetal layers of the cotyledons.

It has been postulated that PL in the maternal circulation plays a role that is advantageous to the fetus by virtue of its lipolytic, anti-glucose utilisation action, thereby sparing glucose which would have been metabolised by the mother for use by the fetus. Whether or not PL plays a role in directing various energy sources in the fetus in times of food deprivation merits investigation.

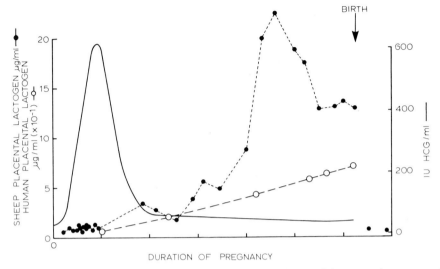

Fig. 8.4. Changes in he concentration of hCG and placental lactogen in maternal plasma in the human and the sheep. Data from Kelly et al. [355].

In addition, the possible role of PL as a growth factor in the fetus remains to be investigated experimentally. It would be of interest to know the concentrations of PL in other body fluids in the sheep fetus and what stimuli, if any, will change these concentrations.

Finally, before leaving this consideration of prolactin and PL, some mention should be made of the suggestion that PL and prolactin may play a role in suppressing maternal immune responses. Since the fetus is a homograft, it is at risk of rejection by the mother. Both prolactin and PL will depress several features of the immune response such as the lymphocyte reaction. Circulating PL and local concentrations of PL at the placental surface may be indispensible to the continued residence of the fetus in the uterus.

We must consider the possibility of an action of fetal hormones on the placenta. There is at least an indirect relationship between the pituitary and the placenta in respect to the control of steroid synthesis (Chapters 10–12 discuss feto–placental interaction). In a recent report of investigations carried out on rat fetuses on the day before birth, male fetuses were heavier than female but placental weights were the same [94]. There was also an indication that fetuses affected the growth of their neighbours. Liggins and Kennedy [395] reported marked vacuolation of the cells of the tips of the chorionic epithelium in their hypophysectomised fetuses. Although they could not correlate the degree of vacuolation with the extent or duration of hypophysectomy, this observation does suggest the possibility of some fetal endocrine control of placental function. Endocrine effects on the placenta may be in part responsible for the sexual dimorphism observed in fetal weight. Placental weights are higher in male rats than female rats. However, other factors are acting as well since the fetal weight: placental weight ratio was higher in the male [94]. Such endocrine factors in the fetus may affect placental function without changing grosser features such as placental weight. Further work should yield important clues as to the control of these important endocrine relationships between fetus and placenta. The hypophysectomised fetal sheep preparation will be of particular use in these studies, since replacement of fetal pituitary hormones is possible singly and in various combinations.

8.5 General conclusions

It is clear that there are several interesting observations regarding the production and the concentrations of these three related hormones within the fetal compartment. Each of these hormones exists in very high concentrations in at least one fetal tissue or body fluid. In the light of the wide number of potential roles they may fulfill it is difficult to view an amniotic fluid concentration of 3×10^{-7} M prolactin as inconsequential. However, at the present time, no definite physiological role can be ascribed to these three polypeptide hormones in the fetus. GH may play a role in amino acid and glucose metabolism either directly or by stimulating insulin release and the changes immediately before birth may demonstrate an interrelationship between the fetal adrenocortical axis and other fetal endocrine systems [562]. A similar interrelationship may exist between the adrenal and thyroid axes. With regard to prolactin it has always been an enigma as to the role of this hormone in the male. Perhaps its major function in the higher mammals is discharged during fetal life.

CHAPTER 9

Adrenocorticotropin

Note on Chapters 9, 10, 11, 12
The final four chapters of this volume are closely interrelated. The present chapter deals with current knowledge and ideas regarding the secretion and control of ACTH in the fetal sheep. The next chapter is concerned with fetal adrenal function, particularly in the sheep and primate fetus. Chapter 11 is concerned with other important factors which probably play a role in the initiation of parturition. The final chapter is an attempt to draw together the various threads in Chapters 9, 10 and 11. The outstanding contribution in this field has been made by Professor G.C. Liggins. As mentioned in the preface, I gratefully acknowledge the stimulus of many hours of discussion with him while working in his laboratory. The reader unfamiliar with this field would be well-advised to read the review article of work from his laboratory [391].

9.1 ACTH in the fetal sheep

9.1.1 Significance of measurements of circulating plasma ACTH concentrations

The estimation of plasma ACTH concentrations has considerable problems even in unstressed, co-operative adult animals. Considerable evidence exists that ACTH is released in response to a wide variety of stressful stimuli [597]. Sampling procedures themselves may stimulate ACTH secretion. In addition, ACTH secretion demonstrates a nyctohemeral or circadian* rhythm in adult

---

* Endocrine rhythms often bear a relationship to the 24-hour daylight cycle. Such relationships are often referred to as circadian (meaning that they recur with a cycle of about one day) or nyctohemeral (meaning that they relate to day and night). It is very necessary to investigate the causative factor directly responsible for this time course. Since so many physiological functions are related to different aspects of the daily cycle,

animals and within that rhythm there are probably spurts of ACTH release [128]. Great caution must be placed on individual estimations of plasma ACTH concentration. However, as with AVP and angiotensin, investigators are placed under the very real constraint that repeated bleeding may of itself constitute a stressful haemorrhagic stimulus to ACTH release.

The assay methodology currently available for ACTH presents several practical problems. There are various pituitary molecules with structural similarities to ACTH (Fig. 9.1). The N-terminal end of the ACTH molecule is the biologically active end and the use of antisera directed against the

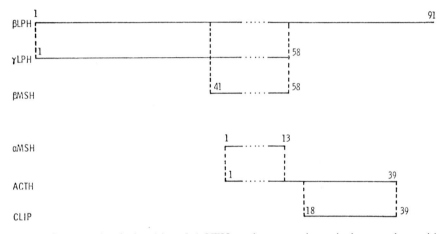

Fig. 9.1. Structural relationship of ACTH and some other pituitary polypeptides. The numbers represent the number of amino acids in each polypeptide. 1 represents the N-terminal end: β-Lipotropic hormone (β-LPH) and possibly γ-lipotropic hormone (γ-LPH) are produced by the pars distalis; α-melanocyte-stimulating hormone (α-MSH), β-melanocyte-stimulating hormone (β-MSH) and corticotropin-like intermediate lobe peptide (CLIP) are thought to be produced by the pars intermedia. The first 58 amino acids of β-LPH are identical in sequence to γ-LPH. β-MSH shares its 18 amino acid sequence with β- and γ-LPH. The first 13 amino acids of ACTH are identical with α-MSH. The 18–39 amino acid sequence of ACTH is identical with CLIP. There is a common heptapeptide core (dashed line) within β-LPH, γ-LPH, α-MSH and ACTH which is responsible for the melanocyte-stimulating activity of these molecules. Reproduced with permission from Professor T. Chard.

various features may be secondary to other functions which are directly related to the changing pattern of daylight and night. It has been demonstrated that feeding affects arterial pCO_2, pH and HCO_3^- concentrations in maternal sheep plasma [127]. It is possible that diurnal rhythms in the fetus are based on 'cues' from such changes. In addition, change of accomodation and confinement may alter circadian rhythms [296].

C-terminal end may give erroneous values as a result of measurement of biologically inactive peptide fragments with the C-terminal configuration. The development of the cytobiochemical assay, based on the biological property of ACTH to deplete ascorbate levels in the steroidogenic cells of the adrenal cortex is a very promising development [120a]. Not only is this assay a bioassay but its sensitivity is such that only very small volumes of plasma are required (of the order of 10 µl), whereas accurate measurement of ACTH by radioimmunoassay requires about 2.0 ml of plasma at the concentrations present in fetal plasma. Repeated fetal sampling is thus possible without the dangers of haemorrhagic stress. Constant attention should be paid to the percentage of fetal blood removed for various analyses. Data is available to permit the calculation of fetal blood volume [91].

9.1.2 Fetal plasma ACTH concentrations in acute experiments

ACTH has been demonstrated in fetal lamb plasma as early as 59 days of gestation [12]. Pituitary concentrations of ACTH show little change between 90 and 143 days gestation. Concentrations at this time are similar to those in the adult. The first observations of fetal plasma ACTH concentrations were obtained using radioimmunoassay methods on plasma samples obtained from acutely exteriorised fetuses. Plasma ACTH concentration was less than 100 pg/ml between 75 and 130 days. In fetuses greater than 130 days, concentrations as high as 1200 pg/ml were observed. It was therefore concluded that, 'in utero there is a rise in ACTH and AVP levels as pregnancy progresses. The alternative possibility, that the stress of exteriorization releases these substances more readily in the older animals, seems less likely since ..., the levels [prior to haemorrhagic or hypoxic stress] remained relatively constant after exteriorisation for periods in excess of an hour' [12].

Thus, these authors quite reasonably attribute the higher resting plasma ACTH concentrations after surgical preparation of fetuses of older gestational age to a higher baseline, undisturbed secretion. If this is so, the higher fetal plasma ACTH concentration reflects the conditions existing in utero as gestation advances. This problem highlights the difficulty of measuring true, undisturbed baseline concentrations. The very act of catheterisation disturbs the existing uterine conditions. The higher resting concentrations in older fetuses is open to a different explanation, namely that older fetuses do respond to exteriorisation with a greater secretion of ACTH than younger fetuses and that these new submaximal levels are maintained throughout the resting period. Further stress in the form of hypoxia or haemorrhage subsequently resulted in further increases in plasma ACTH concentrations during the experiment.

9.1.3 Measurements of plasma ACTH concentrations in chronically catheterised fetal sheep

Several reports of fetal sheep plasma ACTH concentrations are now available. In 5 chronically catheterised fetuses, plasma concentrations measured by radioimmunoassay varied between 100 and 1800 pg/ml over the last 30 days of gestation [15]. Only one fetus for which both plasma ACTH and cortisol concentrations were available for the crucial period of the last ten days of intrauterine life was undisturbed by other experimentation. In this fetus there was no noticeable increase in fetal plasma ACTH concentrations before the sharp increase in fetal plasma cortisol concentrations which occurs prior to delivery.

In a similar study, five fetuses were sampled for 6–14 days before delivery [487]. Four were delivered alive by spontaneous vaginal delivery and one died intra-partum. No increase in fetal plasma ACTH could be observed before the rise in fetal plasma cortisol (Fig. 9.2). The animal which died during delivery

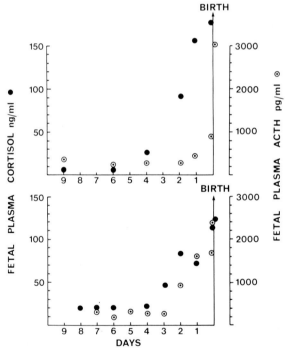

Fig. 9.2. Simultaneous fetal plasma cortisol and ACTH concentrations in two sheep fetuses during the last 9 days of intrauterine life. Normal, healthy, live lambs were delivered at the time indicated by the arrow. Fetal pH and blood gases tensions were within the normal range at the time of each sample. Abscissae, days before birth; ordinates, plasma cortisol (●) and ACTH (⊙) concentrations [487].

Table 9.1.
Plasma ACTH data measured by radioimmunoassay in 5 chronically catheterised fetal sheep. n = number of observations. Mean ± S.E.M. Rees, Jack, Thomas and Nathanielsz unpublished observations.

	12–24 h Post-surgery surgery	Up to 48 h Before delivery	24–48 h Before delivery	0–24 h Before delivery
n	7	44	5	9
Plasma ACTH concentration (pg/ml)	735.7 ± 119.7	365.7 ± 42.1	664.8 ± 147.3	1617.5 ± 257.8

had been sampled every 6 hours for 4 days and this may have been the cause of its death. Other animals were only sampled daily. Table 9.1 shows the data from these animals. Fetal plasma ACTH concentrations were raised in the 24 h immediately following surgery. Thereafter levels remained stable until the last 24 h of gestation when they rose to a mean concentration of 1618 pg/ml. There are two major points of interest with regard to the terminal rise. It is confirmation of the claim that the terminal phase of the increase in fetal plasma cortisol concentration occurs in response to the stress of delivery. This ACTH data is evidence for stimulation of the pituitary adrenal axis at this time. Secondly, it demonstrates that the failure to measure a rise in plasma ACTH preceding the increase in cortisol was not due to a methodological inability to detect increases in plasma ACTH. This applies to both phases 2 and 3 in Fig. 9.3. By itself this failure could be considered to have been due to a methodological inability to demonstrate a rise in fetal ACTH. This is very unlikely since not only was there no increase in fetal ACTH preceding the phase 2 and 3 cortisol increment in any fetus but there was a measurable increase in every animal at the time of delivery. Obviously sequential fetal plasma ACTH concentrations during phase 1 would be of great interest.

There are two further studies of fetal plasma ACTH concentrations which were not directed towards obtaining sequential values. In samples from 26 fetuses of gestational age 96–145 days, values for fetal plasma ACTH ranged from less than 50 pg/ml to around 900 pg/ml. ACTH concentrations did not correlate with fetal blood gas or corticosteroid concentrations and the 17 out of 41 values of more than 200 pg/ml presumably were obtained at various gestational ages since it is stated that 'plasma ACTH concentration was not related to gestational age' [79]. Fetal adrenal growth and the commencement of increasing plasma cortisol concentrations would have occurred by 145 days

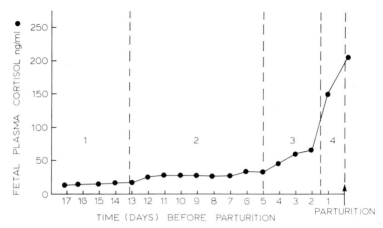

Fig. 9.3. Diagrammatic representation of the rise in fetal plasma cortisol concentrations before delivery in the lamb (see also Chapter 10). As discussed in the text, the changes have been divided into four phases: phase 1, a steady baseline; phase 2, a transition period; phase 3, steady rise in fetal plasma cortisol concentration; phase 4, a rapid pronounced phase of increment in plasma cortisol just prior to delivery.

and yet the fetal plasma ACTH concentrations had not risen significantly in these older fetuses.

Maternal plasma ACTH concentrations in the ewe ranged from less than 10 to just over 100 pg/ml [80, 331]. During phase 2 of the fetal plasma cortisol picture (Fig. 9.3), fetal and maternal total and free cortisol concentrations are of the same magnitude. Indeed the fetal plasma concentration of free cortisol is probably slightly greater than that of the mother yet fetal plasma ACTH concentrations are 10 to 20 times greater than the maternal concentrations. In addition, it appears that the setting of the levels of activity and feedback between cortisol and ACTH may vary considerably between fetuses.

In summary, fetal lamb plasma ACTH concentrations are generally higher than those observed in the pregnant ewe or the adult in other mammalian species [486]. There is no clear evidence that fetal plasma ACTH concentrations are elevated either during phases 2 or 3 of the changes in fetal plasma cortisol concentration described. The rapid growth of the fetal adrenal in the last 15 days or so of gestation (Fig. 9.4) and the increment in fetal plasma cortisol concentration are almost certainly linked together. Several possible causative mechanisms may be put forward to explain these events.

(a) At the present time there is no experimental evidence to rule out the possibility that they reflect an increased secretion of ACTH from the fetal pituitary.

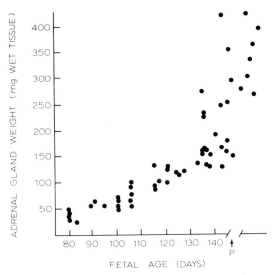

Fig. 9.4. Weight of right adrenal at various gestational ages in normal fetal lambs. P = parturition. Reproduced with permission from Nathanielsz et al. [442].

(b) Other pituitary factors may be of importance.

(c) Adrenal growth and increased cortisol secretion may be produced by an increasing drive from a trophic factor from an alternative source such as the placenta. These possibilities are discussed in detail below.

(d) Increased fetal cortisol secretion may represent an intrinsic maturation of the adrenal gland itself. Such a biological clock may be dependent on a basal level of a trophic stimulus such as ACTH.

9.1.4 Effect of adrenal growth on the increased production of cortisol by the fetal sheep

Surprisingly little attention has been given to the possibility that the increased output of cortisol by the fetal adrenal gland as gestation proceeds simply reflects the growth of the adrenal cortex. We may start our consideration of this question with three fundamental observations of the changes in fetal adrenal activity between 120 days gestation and delivery. Firstly, fetal adrenal weight increases by a factor of 4. In the Welsh Mountain breed the increase in weight is from 100–400 mg (Fig. 9.4). Secondly, in vitro perfusion of fetal adrenal slices composed of medulla and cortex (no separation having been performed) with constant concentrations of ACTH, demonstrate an increased responsiveness to ACTH of about 9.5 times [403] over the last 8 days or so of intrauterine life. Finally, in vivo the fetus demonstrates increased adrenal responsiveness to ACTH infusions over this period.

Fig. 9.5 demonstrates that the size of the fetal adrenal medulla remains reasonably constant throughout this period. Measurements of the adrenals used for these micrographs showed that at 100 days gestation the medulla constituted 62% of the gland whereas at 144 days it was only 38%. The absolute weight of cortex had therefore risen from 38 to a possible maximum of 248 mg – an incre-

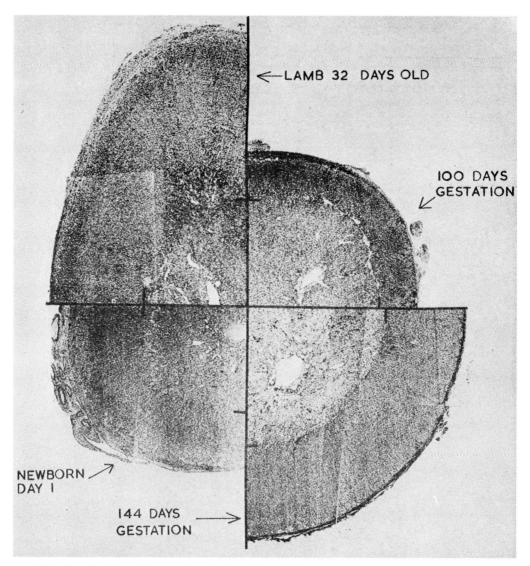

Fig. 9.5. Sections of adrenals of fetal and newborn lambs to demonstrate that the increased fetal adrenal weight is predominantly due to cortical growth.

ment of 6.5 times. Within the limits of experimental error this increase explains two-thirds of the 10-fold increment in fetal cortisol secretion. In the anaesthetised fetal dog, adrenal steroid secretion was more closely related to adrenal weight than gestational age [324].

Even if the basis of the increased adrenal cortical secretion is simply an increase in total adrenal cortical size it is necessary to look for the factor(s) which cause this remarkable growth. Perhaps the fetus in the end obtains the benefit of having an adrenal axis which is set at a high level of ACTH secretion with little negative feedback in the 10–30 ng cortisol/ml plasma range. However, this suggestion poses several further questions. Since fetal plasma ACTH concentrations are high at 110 days gestation why does the spurt in adrenal growth only begin at around 130 days? Are there any differences in the set-point of the feedback in the neonate and how are they brought about? Paisey and Nathanielsz [459a] considered that factors other than a simple hypothalamic resetting must be operating.

9.1.5 Possible diurnal and other phasic changes in fetal ACTH secretion

Some debate exists as to whether the adult sheep demonstrates a diurnal pattern in circulating cortisol concentrations. Two reports suggest that there are increased plasma cortisol concentrations in the adult sheep during the hours of daylight [295, 402]. A third study failed to show a diurnal pattern [41]. One of the problems is that this rhythm as well as others such as feeding rhythms, patterns of glucose and insulin changes, are considerably affected by cage restraint and other abnormal features of experimental conditions. In the adult human the highest plasma cortisol concentrations occur early in the morning whilst in the rat, a nocturnal animal, corticosterone (the major steroid in this species) is present in its highest plasma concentrations in the evening when plasma ACTH concentrations are highest [18, 486]. It appears that ACTH secretion is regulated from the hypothalamus in response to the sleep–wakefulness pattern of the animal. By reversing the feeding pattern of rats and hence their sleep–wakefulness routine it is possible to reverse the peaks and troughs of plasma corticosterone concentrations [371]. It is worthy of note that administration of 500 µg of hydrocortisone acetate (a reasonably large dose) to 2-day-old newborn rats will delay the appearance of circadian rhythmicity in plasma corticosterone concentrations by about 50 days [371, 423]. In addition there is a diurnal variation in the ability of glucocorticoids to suppress adrenal function [449].

At the present time a clear demonstration exists of one circadian rhythm in the fetal lamb. The work of Dawes and his colleagues at Oxford has demonstrated a circadian variation in the incidence of fetal breathing movements in utero [76, 78, 163]. Fetal breathing movements may be monitored by an indwelling electromagnetic flow meter placed in the fetal trachea to monitor fluid flow [165]. Alternatively, the movements of the fetal chest may be observed using ultrasound equipment. Fetal breathing movements can be related to the electroencephalogram (EEG) recorded from electrodes implanted biparietaly in the fetal skull and eye movements recorded from electrodes implanted over the palpebral ridges of the fetus. In such a preparation, fetal breathing movements coincide with the low voltage rapid wave electrocortical activity associated in adult animals with rapid eye movement (REM) sleep.

The fetus normally spends about 40% of the day performing rapid irregular breathing movements. The periods of slow and fast wave EEG activity may last for variable periods, from a few minutes to an hour or more. In an attempt to correlate fetal plasma ACTH concentrations with breathing movements, Boddy and co-workers [79] related the incidence of breathing movements over 1 hour with the plasma ACTH concentration measured either at the beginning or end of the hour (Fig. 9.6) in which the occurrence of breathing movements was measured.

Fig. 9.6 demonstrates that there was an inverse relationship between the incidence of fetal breathing movements and the secretion of ACTH by the fetal pituitary. There is therefore apparently a temporal relationship between REM-sleep, rapid low-voltage EEG activity, fetal breathing and decreased plasma ACTH concentrations.

Other experimental results are available to link these activities. The frequency of respiratory movements is significantly reduced two or three days before parturition [76]. In addition, hypoglycemia leads to a diminution of both amplitude and frequency of fetal breathing. Hypoglycemia has been reported to stimulate the fetal hypothalamo–hypophysial–adrenal system, although others have reported inability to demonstrate a rise in fetal plasma cortisol following fetal hypoglycemia [391, 516]. The explanation of these conflicting observations probably resides in the degree and duration of the hypoglycemic stimulus used. In addition, if measurement of plasma ACTH were an easier task to perform, it would give a clearer indication of changes in the fetal hypothalamo–hypophysial–adrenal axis, since poor sensitivity of the adrenal to ACTH at younger gestational ages blunts the manifestation of any changes in plasma ACTH as changes in fetal cortisol secretion.

It was mentioned in Chapter 8 that GH secretion in the adult occurs predominantly during slow wave sleep. It was also suggested that high fetal GH concentrations may reflect the fact that slow wave EEG activity is present in the fetus for a much larger part of its day than in the adult. Similar considerations may

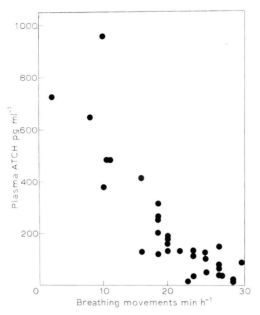

Fig. 9.6. Fetal carotid plasma ACTH concentration and incidence of fetal breathing movements. Each point represents the fetal plasma ACTH concentration either at the beginning or the end of the hour that the incidence of fetal breathing was measured. Observations at both times were made on most animals. Reproduced with permission from [79].

apply to the secretion of ACTH by the sheep fetus. As mentioned above, fetal sheep plasma ACTH concentrations are higher than the maternal concentrations by about the same factor as GH, 10–20 [49]. It is obvious that a search will have to be made for diurnal patterns in fetal hormone concentrations under different physiological conditions. Fig. 9.7 shows some aspects of variability in the plasma cortisol and ACTH concentrations in two sheep fetuses. It is apparent that there are fluctuations in both ACTH and cortisol concentrations. It should be recalled that a diurnal variation in fetal glomerular filtration rate (GFR) has been demonstrated in the lamb [98]. These authors suggest that GFR is being affected by changes in fetal glucocorticoids.

9.1.6 Calculation of fetal ACTH secretion rate

Using a single injection technique with radioactively labelled ACTH, the half-life of ACTH in plasma has been calculated as 1.1 min for the sheep fetus and

1.03 min for the ewe*. The Metabolic Clearance Rate (MCR) for ACTH was 55 ml/min/Kg in the fetus and 34 ml/min/Kg in the ewe. ACTH secretion rates were 4921 pg/min/Kg in the fetus and 1245 pg/min/Kg in the ewe. Plasma ACTH concentration used for these calculations was taken from the mean of 6 measurements by radioimmunoassay over the course of following the disappearance of labelled ACTH. The maximum coefficient of variation for the ACTH measurements was only 28%, which suggests that fetal plasma ACTH concentrations do not fluctuate greatly over a short period of time [331]. However, Fig. 9.7(a) shows that such fluctuation can occur and since the half-life of ACTH in plasma is so short, it would be very easy to miss peaks without sampling at a frequency too great to be tolerated with radioimmunoassay techniques. When very rapid fetal samples are taken and estimated by the cytochemical bioassay, such peaks can be readily demonstrated. One experiment involving the infusion of synthetic ACTH (Synacthen) into three hypophysectomised lamb fetuses around 125 days gestation has been described [330]. Continuous infusion of Synacthen at 10 µg/h resulted in fetal plasma ACTH concentrations of 100–1000 pg/ml. This data can be used to calculate a MCR for ACTH of approximately 1670 or 167 ml/min for the two different concentrations. Assuming a weight of 3.0 Kg for the fetus at this age the MCR becomes 550 or 55 ml/min/Kg, respectively.

9.1.7 Factors other than pituitary ACTH which may affect adrenal secretion of cortisol

Recent measurements of radioimmunoassayable ACTH in pregnant humans has suggested that maternal plasma contains some ACTH which cannot be suppressed by dexamethasone. The placenta is the probable source of this ACTH [241, 485]. Extracts of human placentae contain radioimmunoassayable and bioassayable ACTH even after all the maternal and fetal blood within the placenta has been very carefully washed out. It may be noted at this point that some evidence has also been produced suggesting the existence of a placental TSH [277, 279].

---

* As discussed in the Appendix, Metabolic Clearance Rate (MCR) is best calculated by constant infusion experiments. When a constant concentration [C] of the molecule being infused is reached, MCR is calculated by the formula: MCR = Infusion rate/[C]. It is therefore clear that when ACTH is infused at 10 µg/h (166.6 ng/min), if the final concentration reached is 100 pg/ml, then

$$\text{MCR} = \frac{166.6 \times 1000 \;(\text{pg/min})}{100 \;(\text{pg/ml})} = 1666 \;\text{ml/min}.$$

Similarly, if the final concentration reached is 1000 pg/ml, MCR = 166.6 ml/min. Even the lower value is 3 times that obtained by single injection techniques.

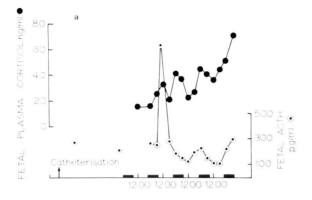
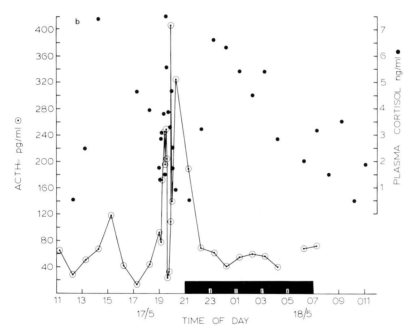

Fig. 9.7.(a) Fetal plasma ACTH concentration measured by radioimmunoassay and fetal plasma cortisol concentration in samples taken every 6 hours from one chronically catheterised sheep fetus. Delivery occurred 24 h after the last sample at 143 days gestational age. The ewe was maintained in a room in which lighting was controlled. Lights on at 07.00 and off at 21.00 as shown by the solid bars. (b) Rapid fluctuations in fetal plasma ACTH concentrations as measured by cytobiochemical bioassay in samples taken over 24 h. The lighting regime was as stated in 7(a). Between 19.00 and 20.00 very small fetal blood samples (0.2 ml) were taken every 5 minutes. (Data from Ratter, Rees, Jack, Thomas and Nathanielsz, unpublished.)

Following fetal hypophysectomy in the sheep at 110 days gestational age, fetal plasma ACTH falls to undetectable concentrations when measured by radioimmunoassay, <10 pg/ml [330] or cytobiochemical assay (<0.34 pg/ml, Ratter et al., unpublished observations). As mentioned in the previous paragraph, in the human there may be a placental ACTH which shares some antigenic properties with pituitary ACTH since it can be measured by radioimmunoassay. It is clear that radioimmunoassay, even with a battery of antisera directed at various locations in the ACTH molecule, cannot exclude the possibility that a molecule of completely different structure may be exerting an action on the developing fetal adrenal and for this reason, the low values in the bioassay are important in attempting to show that all factors which may stimulate the fetal adrenal have been removed by hypophysectomy. One pituitary factor which possibly has actions on the adrenal, prolactin, has already been discussed.

In the light of the inability to demonstrate a rise in radioimmunoassayable fetal plasma ACTH concentrations prior to the rise in fetal plasma cortisol, it is necessary to consider the possibility that another factor with biological adrenocorticotropic activity but immunologically dissimilar to ACTH is responsible for the increase in fetal adrenal size and the stimulation of cortisol secretion. It will become increasingly necessary to use bioassay methods to assess the adrenocorticotropic activity of fetal plasma. In addition, it may prove necessary to use fetal lamb adrenal cells, at a stage when they are responsive, to test biological activity, since it is possible that factors which are trophic to the fetal lamb adrenal may not be trophic in other species. This is very necessary since the search is for other molecules which may stimulate the fetal sheep adrenal at this critical phase. Such a molecule may be missed if adult guinea pigs or rats are used as the source of the adrenal tissue for the bioassay. The hypophysectomised sheep fetus should prove a useful preparation to investigate the possibility of extrapituitary factors which can be shown to have activity in bioassay systems. Similar considerations to those mentioned above apply to attempts at demonstrating the presence of any possible inhibitors of adrenal activity which may be circulating in fetal plasma.

9.1.8 Possible positive feedback systems in the fetal hypothalamo–hypophysial–adrenal axis

Metopirone is a potent inhibitor of adrenal glucocorticoid synthesis. It will block parturition when infused into the fetal lamb. However, infusion of metopirone does not result in fetal adrenal hypertrophy as might be expected in the normal adult animal. There is some evidence to suggest that at fetal plasma cortisol concentrations less than 30–40 ng/ml there is no feedback inhibition of ACTH secretion [322] (Fig. 9.8)*. It is therefore not suprising that metopirone does not stimulate an increased release of ACTH following depression of these low levels of cortisol. One observation of extreme significance from the use of metopirone is that when two fetuses were treated with 100 µg ACTH/24 h in addition to the metopirone, not only was parturition prevented, but so also was the hypertrophy of the fetal adrenal which inevitably follows the infusion of this dose of ACTH when it is infused without metopirone [386]. This observa- shows that as well as blocking the steroidogenic effect of ACTH, metoprine blocked its trophic action. This may be an indirect and generalised effect of interruption of normal metabolism within the steroidogenic cell. Alternatively, since metopirone inhibits the production of cortisol, this effect is compatible with the suggestion that cortisol in some way is necessary or augments the action of ACTH on the adrenal cell.

These experiments should be repeated whilst measurements of fetal cortisol and ACTH concentrations are being made. It has been suggested that the explosive nature of the increase in fetal plasma cortisol before normal parturi- tion, together with the production of a rise in fetal plasma cortisol in the un- infused twin when labour is initiated by the infusion of cortisol into one of a pair twins, are indications of the existence of positive feedback mechanisms affecting the hypothalamo–hypophysial–adrenal axis [442]. The experimental observa- tions with metopirone may be interpreted on the basis that since metopirone inhibits the production of cortisol, cortisol is unable to exert its positive feedback on one or both of two processes, fetal adrenal growth and the stimulation of further cortisol secretion either directly or via further ACTH secretion. Evi- dence that adrenal secretion of cortisol does exert a feedback on ACTH secretion comes from the recent observation that infusion of ACTH at 10 µg/h into three hypophysectomised lamb fetuses produces steady plasma ACTH and cortisol concentrations. Infusion of the same dose of ACTH into intact fetuses produces a swinging pattern of ACTH and cortisol [330]. Plasma cortisol concentrations reached in these animals were about 50 ng/ml. Fig. 9.8 shows that when cortisol is infused into the fetus to produce such concentrations, there is little, if any, negative feedback on the secretion of ACTH by the pitui- tary. It is therefore possible that the fluctuating concentrations of plasma ACTH in the intact fetus simply reflect the normal phasic release of the normal fetal pituitary which is not constrained by any negative feedback. This

---

* There may be other instances in which the setting of feedback levels may differ in the perinatal period. For example, the negative feedback of ovarian steroids may not occur in the newborn rat until day 25 of neonatal life [33].

fluctuation will obviously be lacking in the hypophysectomised fetus. Unfortunately no blood gas data were reported with these experiments and the inevitable absence of other fetal pituitary hormones may complicate simple interpretation. In addition, uterine contraction records are very necessary in experiments involving induction of labour. Stimulation of the fetus by contrac-

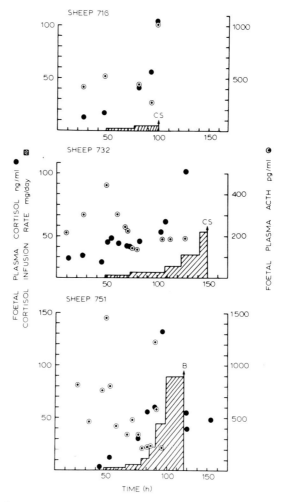

Fig. 9.8. The failure of infusions of cortisol of up to 88 mg/24 h into the fetus to cause significant suppression of fetal plasma ACTH concentrations (⊙) in three fetal lambs. Plasma cortisol concentration (●) and cortisol infusion rate ▨. Fetuses 716 and 732 were delivered by caesarian section in order to obtain tissues for electron microscopy. They were both alive at this time. Fetus 751 was born alive without assistance. Fetal blood gases and pH remained within the normal range throughout the experiment.

tions may affect ACTH release from the intact fetus whilst such a release cannot occur in the hypophysectomised animal. However, taken together with the evidence from the animals given ACTH and metopirone, these experiments suggest that further investigation of the possibility of positive feedback loops in the fetal hypothalamo–hypophysial–adrenal axis is urgently needed.

9.1.9 In vitro studies of changes in adrenal sensitivity to ACTH

As mentioned in Chapter 2 (see Fig. 2.3), superfusion of slices of fetal adrenal glands from fetuses at various gestational ages has demonstrated that the sensitivity of the fetal adrenal cells to a fixed concentration of ACTH increases with gestation. The early insensitivity to ACTH explains the low plasma cortisol concentrations in the face of high plasma ACTH concentrations at earlier gestational ages (100–130 days) compared with high values for plasma cortisol at term. The problem of the explanation of the increase in sensitivity in the apparent absence of a rise in plasma ACTH concentration is discussed above.

9.1.10 In vivo studies of adrenal stimulation by ACTH

Infusion of ACTH (10 µg/h) into fetuses of early gestational age does not produce as good an immediate response in cortisol secretion as it does in the term fetus [47]. However, if the infusion is continued at this rate for 24 h then the adrenal sensitivity increases. Parturition can be initiated in the fetal sheep with infusion rates of 1 µg ACTH/h for 3 days [413]. The infusion of 10 µg ACTH/h therefore probably exceeds the physiological range. This suggestion is borne out by the high plasma corticosteroid concentrations (220 ng/ml), a more rapid rise in fetal cortisol than occurs in vivo at parturition and the high plasma ACTH concentrations (1000–2000 pg/ml) obtained on infusions of 10 µg ACTH/h compared with the usual values at this time, 100 ng cortisol/ml and 200–400 pg ACTH/ml [47, 487]. The possibility of the presence of a positive feedback would in part explain these high values, nevertheless it should be noted that one tenth this rate of ACTH infusion will produce parturition when infused at 126–132 days [413]. Continuous infusion of ACTH at low dose rates will permit the appearance and development of possible positive feedback mechanisms even in younger fetuses (110–130 days), if the various mechanisms, be they hypothalamic, pituitary or adrenal, are ready to respond. Liggins [386] observed that 4 µg ACTH/h was adequate to stimulate delivery when infused into a fetus in 7 out of 9 pregnancies at 88–129 days gestational age. In both the cases in which this dose failed, the pregnancy was a twin pregnancy. It will be recalled

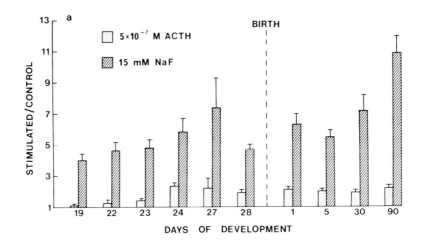

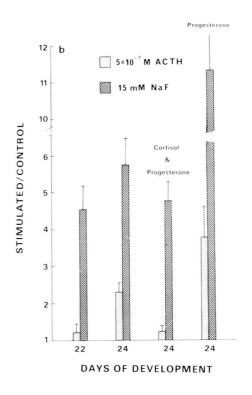

that the data of Jones et al. [331], in which blood production rate of ACTH was calculated from isotope dilution studies, suggest a fetal ACTH production rate of 1.2 µg/h.

9.1.11 Mechanisms of the increase in fetal adrenal sensitivity to ACTH

An increase in the sensitivity of the adrenal to ACTH may constitute the rate-limiting step in the mechanism involved in the initiation of delivery (Chapter 10). As mentioned previously, such an increase in sensitivity is suggested by the in vitro studies and by observing the effects of ACTH infusions in fetuses of different gestational age [47, 403]. Three possible mechanisms to explain this increase in the adrenal sensitivity to ACTH as gestation advances were given above. One of these, that the increased sensitivity simply represents a growth in the size of the adrenal cortex, was discussed. The conclusion that the increase in the amount of cortical tissue may in itself explain the increased cortisol output and the increased adrenal sensitivity was considered only a partial explanation.

As mentioned in Chapter 10, ACTH has its initial effect on the steroidogenic cell by activating the enzyme adenylate cyclase. Increase in the amount of adrenal cortical tissue may not necessarily result in an increase in this complex, receptor and adenylate cyclase, in a functional state.

Further evidence for an interrelationship of cortisol and ACTH in control of fetal adrenal development has been obtained in the fetal rabbit. In this species, adrenal ACTH-stimulatable adenylate cyclase activity increases dramatically between 22 and 24 days gestation. When 0.3 mg cortisol/h is administered to the pregnant doe for 3 days from 21 to 24 days gestation, both the development of ACTH-stimulatable adenylate cyclase activity and normal morphological development of the fetal adrenal were suppressed (Fig. 9.9).

◀ Fig. 9.9. (a) Development of adenylate cyclase activity in the fetal and newborn rabbit. Adrenals were incubated with medium, medium plus 15 mM sodium fluoride (NaF), a non-specific stimulus to adenylate cyclase, or medium plus 5×10^{-7} M ACTH. There is a marked increase in the ability of ACTH to stimulate cyclic AMP production betwee 22 and 24 days of fetal life. (b) The effect of cortisol infusion for 2 days from days 22–24. Cyclic AMP production is plotted as a ratio to that of the same adrenals in medium alone to remove between-animal variation. The cortisol-infused animals were injected with 2.0 mg progesterone intramuscularly to prevent premature delivery. The suppression of specific ACTH-stimulated adenylate cyclase activity could not have been produced by the progesterone, since in the progesterone-injected controls, cyclic AMP production was in fact enhanced. The depression of ACTH effect by the infusion of cortisol to the doe would therefore have to occur in the presence of this enhancing effect of progesterone [6, 319]

Cortisol will cross the rabbit placenta in the maternal–fetal direction. The dose of cortisol administered to the does was intentionally extremely high in order to achieve cortisol concentrations in the fetus which were likely to suppress fetal ACTH secretion. Fetal plasma ACTH concentrations were not measured for the practical reasons of available plasma volumes discussed previously. If fetal ACTH was indeed suppressed, then these observations suggest that ACTH plays a role in the induction of the receptor subunit of the adenylate cyclase receptor–enzyme complex. This would explain the increase in ACTH sensitivity which follows the infusion of ACTH into the sheep fetus. It should be noted that these results are from a polytocous species and extrapolation to the sheep should not be undertaken too rapidly. In addition, these experiments should be repeated with measurements of fetal plasma ACTH concentrations, since the experiments reported above involving the fetal lamb pituitary–adrenal axis suggest that ACTH secretion is not suppressible with physiological plasma concentrations of cortisol. It should be noted that the fetal rabbit has very low fetal plasma cortisol concentrations, approximately 2.4 ng/ml between 27–32 days gestation (Jack, Malinowska and Nathanielsz, unpublished observations using Sephadex LH20 chromatography and competitive protein binding).

Continuous infusion of dexamethasone for several days at 1 mg/day into the sheep fetus will increase the sensitivity of the fetal adrenal to 1-h infusions of ACTH (10 µg/h) into the fetal sheep (Liggins, personal communication). Dexamethasone may be acting via the hypothalamus or pituitary, possibly in a manner similar to that suggested as the explanation of the apparent feedback of cortisol on ACTH secretion in intact fetuses infused with a steady level of ACTH [330]. The apparent contradiction of the negative effects of high concentrations of cortisol on fetal rabbit adrenal development, and this data from the lamb, serves to demonstrate the importance of rigorous consideration of dose–response effects within one species at physiological and pharmacological concentrations, preferably using the physiological glucocorticoid, cortisol.

An alternative explanation is that glucocorticoids may be a factor in the control of development of the receptor subunit on the adrenal cell surface. The observation that metopirone blocks the stimulation of adrenal growth following the administration of exogenous ACTH is in keeping with the possibility that cortisol has a role to play in the development of the adrenal ACTH receptor subunit. A diagrammatic scheme of possible interactions in this system, compatible with currently available experimental results, is shown in Fig. 9.10. Other possibilities exist in addition to those shown here. Cortisol is known to have effects on pituitary function, and in the adult it is possible that there are shortloop feedback effects of ACTH on hypothalamic function [428].

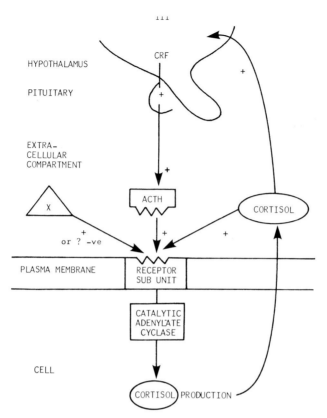

Fig. 9.10. Diagrammatic representation of possible factors which influence the development of the receptor subunit of ACTH-sensitive adenylate cyclase in the fetal adrenal glucocorticoid-secreting cell. +, stimulatory factors; -ve, inhibitory. CRF = corticotropin-releasing factor. X includes all other possible factors of whatever source, e.g. placental.

It would be of interest to know the endogenous plasma cortisol and ACTH concentrations in fetal sheep treated with doses of metopirone that inhibit the stimulatory action of ACTH on fetal adrenal development. It would also be important to assess the effect of fetal adrenalectomy on fetal plasma ACTH concentrations. Although it appears that there is no appreciable degree of negative feedback of cortisol at the low fetal plasma cortisol concentrations in phase 1 and 2, it remains to be seen whether the postulated positive feedback of cortisol does exist (Fig. 9.3).

As discussed in Chapter 12, the extensive studies on the induction of delivery in the fetal sheep preparation have involved the use of a wide array of different agents acting on the adrenal axis. These agents have been given at very many doses, by several routes and in different regimes of single injection or continuous infusion. It is likely that cortisol, which is one of the major factors, acts at several levels of the axis and may have both positive and negative actions of its own and interactions with ACTH. Fig. 9.10 shows some possible interactions between a relatively static circulating concentration of ACTH in fetal plasma and a slowly increasing circulating concentration of cortisol (or even a local concentration) which could produce a positive feedback effect on both the development of the ACTH receptor and the general growth of the steroidogenic cell. With such a system there would not need to be an increment in fetal pituitary ACTH secretion during phases 1, 2 and 3. There is need for further studies of the various endocrine changes produced by these different treatments.

Comprehensive dose–response curves are lacking for dexamethasone or cortisol administration to the fetus other than the original observations of Liggins. Some interesting data is available in the cow. It is not intended to discuss this material here since it will be reviewed elsewhere [440]. However, it clearly shows that dexamethasone infusions to the fetal calf at rates from 0.1–10 mg dexamethasone/day will produce very different endocrine patterns in mother and fetus, depending on the dose rate [Fairclough and Welch, personal communication; 440].

9.2 Function of the fetal hypothalamus

Section of the fetal lamb's pituitary stalk before 116 days gestation will prevent parturition at term. Such fetuses must be delivered by caesarian section [391]. Stalk section at 116 days or after leads to premature delivery. Prematurely delivered lambs had enlarged adrenals which suggested that the surgical act of cutting the pituitary stalk and the insertion of a silicone plate stimulates the secretion of ACTH. The problems of fetal pituitary stalk section with the risk of subsequent infarction are discussed in Chapter 8 in relation to the regulation of GH secretion.

There is a need for data on circulating fetal plasma ACTH and glucocorticoid following stalk section at different gestational ages. It has been demonstrated that complete isolation of the pituitary from its hypothalamic connections in both the adult dog and monkey does not result in decreased adrenal secretion of glucocorticoid over periods of up to 5 days after surgery [357]. Indeed secretion may even be increased over this period of time.

Jost, Dupouy and Rieutort [342] using the encephalectomy technique have produced results in the fetal rat which suggest that the removal of hypothalamic influences at day 19 of gestation results in decreased adrenal growth by day 21.

There was no difference in the adrenals of encephalectomised and decapitated fetuses. In the latter preparation, the pituitary is removed as well. However, the adrenal weights of fetuses encephalectomised at 17 days gestation and harvested at 19 days were larger than those of animals decapitated at 17 days but still significantly less than in control fetuses. Further evidence of a hypothalamic effect was obtained by the administration of corticotropin-releasing factor (CRF) to encephalectomised fetuses. CRF restored the adrenal growth to normal. The conclusion from these experiments was that although removal of hypothalamic influences produces a decrease in function, there is some degree of autonomous activity of the fetal pituitary with respect to ACTH secretion. However, in the older fetuses the responsiveness of the pituitary to the negative feedback exerted by corticosterone from the mother has developed.

Independent pituitary ACTH activity could be observed from the continued adrenal growth in five lamb fetuses in which the pituitary stalk was sectioned at 109 days gestation or earlier [391]. Mean adrenal weight at 160 days was 527 mg, whereas spontaneously delivering lambs in which stalk section was performed at 116 days or later had adrenals of mean weight 619 mg. This difference is not very large and it is of great interest that in one fetus of a pair of twins, both of which were stalk sectioned, the adrenal was the largest of all ten fetuses (Table 9.2). Attention has been drawn to other suggestions of inhibitory phenomena in twin pregnancies, and here it seems that the presence of the very

Table 9.2.
The effects on duration of pregnancy of section of the fetal pituitary stalk. Reproduced with permission from Liggins et al. [391].

Sheep no.	Mode of delivery	Maturity at operation (days)	Maturity at delivery (days)	Adrenal weight (gm)	Body weight (gm)
880	CS[a]	88	160	350	6675
746	CS	97	160	850[b]	6850
				420	6750
584	CS	106	160	635	7400
322	CS	109	160	380	5850
959	Spontaneous	116	124	580	3775
527	Spontaneous	122	142	615	3000
978	Spontaneous	130	141	542	2400
898	Spontaneous	131	145	720	2350
504	Spontaneous	133	140	640	2975

[a] CS = Caesarean section.
[b] Twins.

large adrenal was insufficient to bring about delivery. In addition, another animal (No. 584) had adrenals as large as all but one of the spontaneously delivering fetuses. Possible pituitary factors, other than ACTH, which affect delivery, are discussed elsewhere (p. 197) but this observation also suggests that normal adrenal growth to the critical size necessary for delivery in the normal animal may not be adequate in animals in which fetal pituitary function has been altered. It is possible that some other pituitary factor has an action on the various mechanisms which must be set in motion to initiate the process of delivery.

In summary, it appears that there is some indirect evidence for independent ACTH secretory activity of the pituitary in the adult dog and monkey and the fetal rat and sheep. Measurement of ACTH and corticosteroid concentrations are required to assess directly whether this independent function exists in the fetus and what, if any, is its extent. A high level of activity in the fetal pituitary–adrenal axis in stalk-sectioned fetuses, if it exists, would argue against fetal ACTH being the only trigger to delivery, since stalk section prevents parturition.

9.2.2 Responses of the adrenal axis to various forms of stress

Stresses other than the stress of delivery have been shown to elevate fetal plasma ACTH concentration in the chronically catheterised fetal lamb. Hypoxia or fetal haemorrhage will cause a pronounced secretion of ACTH by the fetal pituitary as early as 96 days of gestation [17, 80]. Fetal plasma ACTH concentrations increase to levels of over 1000 pg/ml. These concentrations are capable of producing parturition when ACTH is infused into the sheep fetus [330]. It should be noted that adrenal blood flow may increase by a factor of 3 during fetal hypoxia [131]. Such vasodilation will expose the fetal adrenal cells to a greater number of ACTH molecules per unit time, even without an increase in in fetal plasma ACTH concentrations.

However, before about 135 days of gestation the fetal lamb adrenal only responds very slightly to these levels of ACTH. Experiments in which ACTH has been infused into the sheep fetus to increase adrenal responsiveness show that in order that a particular stress should produce a significant degree of cortisol secretion by the fetal adrenal, the increased pituitary ACTH secretion must continue for several hours. It is therefore possible that chronic stress to the fetus may produce premature adrenal growth.

As mentioned in Chapter 2, the fetal sheep preparation is not suited to the single-point analysis of tissue concentrations of hormones. In addition, gland content of hormones is difficult to interpret in terms of overall, efficient activity of the system under investigation. Consideration of activity of hypothalamic releasing hormones produces several problems. These small polypeptides have very short half-lives in peripheral plasma, probably due to enzymatic degradation. At the present time it is very uncertain whether assays for hypothalamic releasing factors in peripheral blood will be of routine use in assessing the level of activity of the hypothalamic neurones controlling individual adenohypophysial hormones.

For these reasons, no useful comment can be made regarding the activity of CRF in the sheep fetus. Obviously clear indications of the level of this hypothalamic activity are needed. One study conducted in the rat, does refer to CRF activity in the hypothalamus. In an investigation mainly centred on the development of stress-responsiveness of the adrenal axis in the neonatal period, hypothalamic CRF activity was also measured in fetuses of unstated gestational age. The fetus was removed from the uterus and then exposed to ether stress. Hypothalamic CRF activity was increased by the stress. Corticosterone secretion was also increased but the response was sluggish compared with that in the neonate. This sluggish response was considered to be a possible indication of inadequate function of the portal system [288].

Before leaving our consideration of fetal hypothalamo–pituitary function, two points which are discussed elsewhere should be recalled. Firstly, it is possible that AVP plays a role as a physiological CRF [272]. In their review of this subject, Yates and Marran write – 'Although the available evidence leaves no doubt that vasopressin is a CRF, it seems almost certain that it is not the hypothalamic CRF... Various investigators have had great difficulty agreeing which stress responses are diminished by neurohypophysectomy. However, the important point is that neurohypophysectomy, like congenital diabetes insipidus, fails to prevent the ACTH release response to a variety of stimuli; therefore, vasopressin cannot be CRF' [597]. A note of caution should be added in respect to the particular situation of the fetus. As long as there can 'be no doubt that vasopressin is a CRF' there is the possibility that in the fetal situation it is active as such. As discussed before, the fact that fetal plasma AVP and ACTH concentrations do not appear to be related in their time courses at the time of delivery does not rule out the possibility that AVP is responsible for increases in plasma ACTH at this time.

The final point is concerned with the negative feedback of glucocorticoids on ACTH secretion. In the adult animal, negative feedback of glucocorticoids occurs at several sites, both the pituitary and higher neural levels. It is also clear that the rate at which steroid concentrations change is an important factor in the feedback mechanism [597]. With respect to the sheep fetus at the present time all that can be said is that exogenous administration of cortisol at rates which increase plasma cortisol concentration to 200 ng/ml are compatible with the continued presence of ACTH in fetal plasma. Fig. 9.9 demonstrates that cortisol concentrations of more than 100 ng/ml are accompanied by very little depression of plasma ACTH concentration. In the rat fetus it appears that dexamethasone and cortisol feed back at the pituitary and hypothalamic levels [178].

CHAPTER 10

The production and role of glucorticoids from the fetal adrenal gland

10.1 Glucocorticoid biosynthesis in the adrenal gland

The basic structure of the steroid molecule and the interrelationship of the different groups of steroids have been discussed elsewhere (androgens, Chapter 3, Fig. 3.1; estrogens, Chapter 11, Fig. 11.1). The more important enzymes involved in the synthesis of the two major glucocorticoids, cortisol and corticosterone are shown in Fig. 10.1. The adult adrenal in all mammalian species can carry out all the enzyme steps shown. In contrast, activity of these enzymes may not be fully expressed in the fetus. The consequence of deficiency of action of individual enzymes are discussed below in relation to the concept of inter-

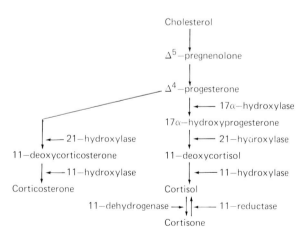

Fig. 10.1. The major biosynthetic steps and enzymes involved in the synthesis of cortisol and corticosterone.

action of the placenta and the fetus. The major glucocorticoid secreted by the adult sheep and human is cortisol [16, 42, 74, 121].

10.1.1 Action of ACTH on the adult adrenal

The chemistry and physiological actions of ACTH have been reviewed extensively elsewhere [293, 294, 561]. ACTH shows inter-species variation, as do all the pituitary polypeptide hormones. It is basically a straight chain polypeptide containing 39 amino acids. The biologically active portion of the molecule resides in the 24 amino acids at the N-terminal end. However, smaller portions of the sequence have been synthesised and do have some biological activity. Several other molecules are present in fetal and adult tissues which have structures similar to portions of the ACTH molecule. Thus, α-MSH is composed of the first 13 amino acids of the ACTH molecule. Little is known of the function of α-MSH, which is present in the fetal human pituitary but not normally in the adult.

ACTH is secreted from cells of the adenohypophysis which show different staining characteristics under different physiological and pathological conditions. Thus, they are generally chromophobe cells which do not stain with the usual stains used for light microscopy of the pituitary. The granules do, however, have a distinctive appearance on electron microscopy and can be localised by immunofluorescence techniques. ACTH secretion is controlled by the hypothalamus. Hypothalamic extracts which will stimulate the release of ACTH both in vivo and in vitro have been prepared but the exact structure of the releasing factor has not been defined. It is probably a small polypeptide like TRH and luteinising hormone-releasing hormone (LH-RH), both of which have been chemically synthesised.

ACTH acts on the adrenal cell by binding to a receptor protein on the cell surface. This receptor is thought to be a portion of the adenylate cyclase complex (see Fig. 9.10). Attachment of ACTH to the receptor activates the catalytic portion of the molecule. Adenylate cyclase catalyses the formation of cyclic AMP from ATP. This begins a chain reaction of enzyme steps within the adrenal cell which result in synthesis and subsequent secretion of steroids from the cell. In the adult, the predominant glucocorticoid is cortisol in most species. Some species, such as the rat, secrete corticosterone. Recently it has been suggested that a slightly different form of ACTH, intermediate ACTH, will act on adrenal cells to stimulate predominantly the secretion of corticosterone [142]. This observation may be of interest in certain phases of fetal development. It has been demonstrated that ACTH will stimulate the 17α-hydroxylase in the rabbit adrenal to increase the ratio of cortisol:corticosterone secreted by the gland [236]. Finally, it should also be borne in mind that ACTH does have several well-documented extra-adrenal effects in adult mammals [561]. It is interesting to recall that ACTH is present in lampreys and other primitive organisms which do not possess an adrenal gland.

10.2 Development of function in the fetal sheep adrenal cortex

10.2.1 Growth of the fetal adrenal

One of the most remarkable growth patterns exhibited by any fetal organ is that of the fetal lamb adrenal. The increase in weight appears to occur in two very distinct phases (Fig. 9.4). Between 80 and 135 days there is a slow, steady growth rate. This is followed by an explosive period of growth during the last 10–15 days of fetal life. This increase in weight is almost entirely due to an increase in the size of the adrenal cortex (Fig. 9.5). Within the cortex the major growth occurs in the zona fasciculata.

Electrocoagulation of the fetal pituitary as early as 93 days gestation results in hypoplasia of the fetal adrenal cortex [392, 395, 442] and a fall in fetal plasma corticosteroid concentrations [442]. The histological changes following hypophysectomy are mainly in the zona fasciculata. Infusion of ACTH into the fetus will reverse the deficiencies of hypophysectomy. It would therefore appear that some pituitary factor (or factors) is responsible for both phases of growth of the fetal adrenal. ACTH is the pituitary hormone most likely to be involved in these changes occurring in the fetal adrenal cortex (see Chapter 9).

10.2.2 In vitro studies of fetal glucocorticoid secretion

The fetal lamb adrenal secretes both corticosterone and cortisol [403]. When adrenal glands taken from fetuses less than 130 days are incubated in vitro, the ratio of cortisol: corticosterone secreted is much lower than that from adrenal glands taken from fetuses within a few days of term. As mentioned previously, in vitro studies have shown that the sensitivity of the fetal adrenal to ACTH increases as term approaches, and that the ratio of cortisol: corticosterone in the secretion also increases. The conclusion from these observations of an increased cortisol: corticosterone ratio following either ACTH treatment in vivo or during normal maturation, is that the activity of the 17α-hydroxylase enzyme increases in both of these situations. It has also been suggested that the immature fetal lamb adrenal has a decreased activity of 11β-hydroxylase in addition to deficient activity of 17α-hydroxylase [23, 24]. However, other in vitro studies have suggested that the fetal lamb adrenal from 100 days gestation to term, does not have a specific enzyme deficiency [159]. In summary, in vitro and in vivo studies demonstrate both an increased activity of the adrenal cortex, an increased sensitivity to ACTH and an increase in the secretion of cortisol as term approaches.

An exact understanding of the development of sensitivity of the fetal adrenal to ACTH, together with the changes in the secretory pattern of the adrenal, is fundamental to the study of the fetal adrenal and the role it plays both during gestation and at parturition. In one study of adrenocortical function from 40 days gestation, fetal plasma cortisol, corticosterone and aldosterone concentrations were higher at 60–90 days gestation than at 91–120 days gestation. Whereas there have been several investigations of hormone concentrations in fetal lamb plasma during the last 30 days of gestation, this useful study is unique in that measurements have been made very early in gestation [592]. The observation of high steroid concentrations early in gestation is remarkable and should be seen in the context of the findings that early in gestation the in vitro production of cortisol was greater per unit weight of tissue than at any other time, except in adrenals taken from fetuses of ewes that were in labour.

The effect of incubation of adrenals in vitro with 1.25 U ACTH/ml (1.25 × 10^9 pg/ml of Synacthen) was also investigated. The ACTH-stimulated output of cortisol was also greater at this early period, 40–49 days, than at any other time except in adrenals taken from fetuses of ewes in labour. When cortisol output was expressed as output relative to the body weight of the fetus, the output from adrenals taken at 40–49 days was even greater than adrenals from term lambs of ewes in labour. There are therefore three different indications of increased activity of the fetal lamb adrenal at 40–49 days gestation, high plasma cortisol concentrations, high resting in vitro cortisol secretion and a marked response to ACTH.

The conclusions to be drawn from these observations are that after a period of activity around 40–50 days, the fetal adrenal becomes relatively quiescent and partially loses its sensitivity to ACTH. ACTH has been demonstrated in the plasma of exteriorised sheep fetuses as early as 59 days gestation [13]. Although changes in plasma ACTH have been demonstrated in the fetal lamb just before delivery (see below), as discussed in the previous chapter, in the period from 130 days until the rise during labour, fetal plasma ACTH appears to remain steady [14, 487]. Sequential studies have not, however, been performed between 49 and 120 days. It remains to be seen whether the decrease in adrenal responsiveness is due to a decrease in plasma ACTH or some other trophic factor. An alternative mechanism would be the appearance of an inhibitor of adrenal activity.

It is of interest that a similar, and possibly related phenomenon of decrease

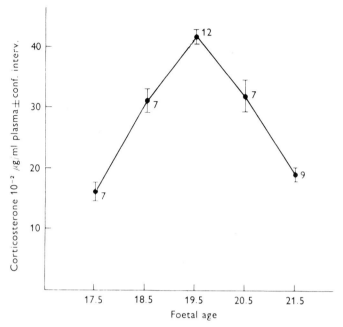

Fig. 10.2. Plasma corticosterone concentrations in the fetal rat. Data from A. Cohen. Redrawn and reproduced with permission from Jost [338].

in adrenal activity has been noted in fetal rat adrenals studied in vitro [501]. Corticosterone (the major glucocorticoid in plasma in this species) was observed in the incubation medium when fetal adrenals removed at 13.5–15.5 days were incubated in vitro. From 15.5 to 19 days no corticosterone was produced. After 19 days corticosterone production reappeared. These changes may have been related to the rapid adrenal growth which occurred after the 16th day [333, 337]. In this report it is also stated that no ACTH-induced stimulation occurred in vitro at any age.

These in vitro results must be interpreted in the light of studies on plasma corticosterone concentrations in the fetal rat. Fig. 10.2 shows that maximal concentrations are not present until 19.5 days but although concentrations are lower before day 19, corticosterone is present in plasma. Other studies have demonstrated secretion of corticosterone into fetal rat plasma as early as 16 days gestation [72, 179, 180, 189]. The situation is complicated in the rat by the fact that the placenta is relatively permeable to corticosterone. It is not clear what proportion of fetal plasma corticosterone is of fetal origin. Measurement of fetal plasma corticosterone following maternal adrenalectomy demonstrates that the fetal adrenal can secrete corticosterone in vivo. However, this experimental system may not reflect in vivo secretion in the undisturbed, normal state.*

10.2.3 In vivo studies on plasma corticosteroid concentrations in late gestation

Since the original observation by Bassett and Thorburn [46], of a rise in fetal plasma corticosteroid concentrations in the period immediately before birth, similar demonstrations of increased fetal plasma glucocorticoid concentrations have been reported by five further groups of workers (Table 10.1). There is an approximately 5–6-fold increase in the concentration of corticosteroid between day (-10) and day (-2) of intrauterine life. One important study in the mouse confirms placental transport of corticosterone in the rodent, and suggests that tissue distribution of corticosterone may not be the same in the mother and the fetus [600]. Within the last few hours before delivery there is a further sharp rise in fetal plasma corticosteroid concentrations. At this time concentrations may increase to greater than 200 ng/ml. This terminal rise almost certainly reflects the response of the fetus to the stresses imposed by the final stages of labour. It should however be recalled that any fetal blood gas changes usually occur very late in the processes of parturition and therefore anoxia is probably not the cause of this cortisol rise. Although the exact stimuli to this release of cortisol are not known it is likely that they are mechanical. In addition, metabolite and respiratory gas changes may also play a role in individual fetuses.

In the first few minutes after delivery a further increase in plasma cortisol often takes place. The stimuli to this increase are probably multiple. They include changes in temperature, humidity differences and tactile stimuli. There may also be rapid, short-lived changes in blood gas composition as the new-

---

* There is considerable agreement regarding the time course and proportion of the corticosteroid increment in fetal plasma. It is, however, certain that different methodologies will produce different results in terms of the absolute concentrations measured. The only comparison of two methodologies is that of Drost et al. [175]. It has been observed by one group of workers that fetal hypophysectomy decreases adrenal glucocorticoid secretion [442]. Two other reports state that adrenalectomy has no effect on the basal concentrations of glucocorticoid [39, 330]. It has also been claimed that bilateral fetal adrenalectomy does not affect circulating glucocorticoid concentrations in the fetal sheep [39]. As discussed below, increasing attention is being focused on the early phase of the fetal plasma corticosteroid rise (around 10 days before delivery). It is therefore very important that the origin and chemical nature of the corticosteroids measured at this time are clearly defined. It is to be hoped that the use of more specific antisera and detailed comparisons of methodology will achieve this.

Table 10.1.
Observations on the increase in fetal plasma corticosteroid concentrations in chronically catheterised fetuses on day −10 and on day −2 (24–48 h before delivery). Note that some of these values have been read off graphical representation of mean values in the literature and are therefore only accurate to the nearest 10 ng/ml. The data of Drost et al. [175] contains a comparison of competitive protein binding (CPB) and fluorimetric methodologies without any chromatographic separation step.

Method	Plasma corticosteroid concentration (ng/ml)		Reference
	Day (−10)	Day (−2)	
CPB, no separation	20	70	46
CPB, no separation	30	100	135
CPB, no separation	10	60	15
CPB, no separation	10	40	175
Fluorimetry, no separation	20	90	
CPB, with Sephadex LH 20 column chromatography	7.5	70	391
Radioimmunoassay with TLC separation	7	40	539

born animal establishes normal, rhythmic and efficient respiration. These changes in plasma cortisol which occur in the first minutes of extrauterine life demonstrate the danger of extrapolating from hormone concentrations observed in neonatal blood to conditions which existed in utero, even at late stages of labour [405, 503]. The role of ACTH in these changes will be discussed later.

The terminal rise in fetal plasma cortisol may play a role in protecting the fetus from the stresses of delivery. This suggestion arises from experiments in which one twin of a pair of intact, catheterised lamb fetuses is infused intravascularly with a high enough dose of cortisol to precipitate delivery at 125–135 days gestation (see below). The uninfused twin usually dies in the last few hours of gestation. In contrast, the infused lamb is born alive and if its maturity is more than 135 days gestational age, it will generally survive and thrive. The mechanism whereby cortisol protects the fetus during delivery is not well understood. Cortisol is concerned with the mobilisation of carbohydrate reserves to maintain blood sugar concentrations. It also has interactions with several other endocrine systems, particularly the thyroid axis (see Chapters 5 and 6). These mechanisms almost certainly have survival value during delivery. In addition, cortisol will affect survival of the newborn animal after delivery, by processes which may involve some of these same mechanisms. However, both before and after birth one of the most important requirements is that for adequate pulmonary surfactant (see p. 165). Cortisol has been clearly demonstrated to play a role in the production of surfactant. However, cortisol requires around 36–48 hours to induce the various enzymes responsible for the synthesis of surfactant. Therefore, the terminal rise in fetal plasma cortisol, phase 4, will not have an adequate time to stimulate further surfactant production. The cortisol in phases 1 and 2 is the important factor in this production.

10.2.4 Changes in fetal plasma corticosteroid concentrations at early stages of gestation

From 60 to about 135 days gestation the pattern of circulating plasma corticosteroids suggests an immature and relatively inactive fetal adrenal (Fig. 10.3) [591, 592]. The possibility of a decrease in fetal adrenal secretion of cortisol for a period after 50 days has been mentioned. In addition, between 90 and 120 days the ratio of plasma cortisol concentrations to that of the 11-deoxy precursors of cortisol and corticosterone is lower than at any other time.

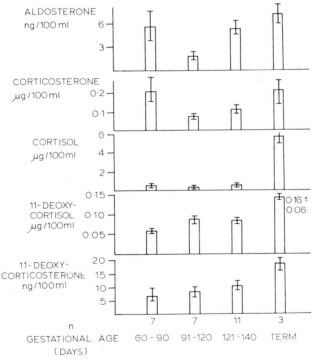

Fig. 10.3. Peripheral blood concentrations of adrenal steroids in the fetal lamb in blood taken under maternal spinal anaesthesia at 60 days to term. Data from Wintour et al. [592].

A most important question relating to the development of the fetal adrenal is when does the pre-parturient increment in fetal plasma cortisol begin? This question was briefly touched upon in the previous chapter when the values for fetal plasma corticosteroid concentrations were arbitrarily divided into four phases (Fig. 9.3). It is at present impossible to answer this question for four reasons. The maximum number of fetuses used in any one study reported to date is 10 and the problems of measurement of cortisol at the 5–20 ng/ml level would require many more catheterised fetuses to establish the first indication of a significant change. Secondly, all investigations reported at the present time have been based on daily, or occasionally twice daily, fetal sampling. The possibility of diurnal endocrine patterns in the fetus (see below) indicates that it may be necessary to sample more frequently at the critical period. Thirdly, changes are occurring in the cortisol-binding proteins at this time and individual measurements of dialysable cortisol would be necessary to define the timing of the physiologically meaningful change. Finally, the problems of assay methodology mentioned in the footnote on page 126 suggest that closer scrutiny must be given to the nature of the molecules being measured.

10.2.5 Placental permeability to cortisol

Assessment of fetal adrenal cortisol secretion in relation to circulating fetal plasma cortisol concentrations requires investigation of placental permeability to cortisol. In the unanaesthetized pregnant ewe the placenta possesses a considerable diffusion resistance to cortisol [61, 172, 391]. This may not be true under general anaesthesia [442].

The most complete data for blood production rates and transplacental transfer rates of cortisol are those of Liggins et al. [391]. The data was obtained using isotope dilution techniques with the continuous infusion of [^{3}H]cortisol into either the fetus or ewe on separate occasions. Metabolic clearance rate (MCR) from the infused pool is calculated from a knowledge of the plasma concentration of radioactivity when equilibrium is achieved (Appendix). The use of radioactive isotopes to study the dynamics of endocrine secretion has been employed most effectively in relation to the study of steroids. MCR, usually expressed in litres/min, is that theoretical volume of plasma which would have to be completely cleared of the hormone per minute to account for the quantity of hormone which has been cleared from the hormone pool (see Appendix). It should not be automatically assumed that these hormone molecules have been active.

From the data in Table 10.2 it will be seen that for a maternal: fetal concentration gradient of about 5 (plasma concentrations of 14 and 3 ng/ml) at day 19, the total maternal transfer of cortisol to the fetus is 0.16 mg/24 h. In the author's experience infusions of cortisol into the fetus at a rate of 5.6 mg/24 h for 7 days are not quite enough to produce delivery; this is greater than 35 times the transfer of cortisol from ewe to fetus at a gradient of 5:1. Thus, it would appear that a transplacental gradient of 175:1 would not result in sufficient transfer of cortisol from ewe to fetus to initiate delivery. For a fetal plasma concentration of 3.0 ng/ml, maternal plasma cortisol would have to reach about 500 ng/ml to produce the required transplacental gradient. Even under the stress of general anaesthesia and abdominal surgery, maternal cortisol concentrations rarely rise above 100 ng/ml (to give a maternal:fetal ratio of about 20) and then only for short periods which are probably inadequate to initiate delivery. Although these calculations do not allow for increases in the proportion of maternal cortisol that is non-protein-bound as the total maternal cortisol increases, it is unlikely that the proportion of dialysable cortisol in maternal plasma will increase sufficiently to bring the ratio to 175.

Liggins [386] was unable to precipitate delivery at 114 days gestation following the infusion of 50 mg cortisol/day for 120 h and 100 mg cortisol/day for a further 72 h into the ewe. The normal BPR for cortisol in the pregnant ewe is about 15 mg/day [56]. It therefore appears that a 7-fold increase in maternal BPR for cortisol is inadequate to produce delivery.

It has been suggested that elevated maternal plasma cortisol concentrations could precipitate delivery [115]. The above calculations show that this is unlikely. If parturition follows maternal stress such as surgery it is likely to be due to stimulation of the fetal hypothalamo–hypophysial–adrenal axis. This is very probable following surgery in mature fetuses in which all the links in the system are active.

10.2.6 Changes in cortisol-binding proteins and cortisol BPR in fetal plasma during the last days of gestation

Fig. 10.4 shows that there is an increase in transcortin binding capacity in fetal plasma over the last 10–15 days of fetal life. In two fetuses the unbound, dialysable cortisol increased from 0.24 to 2.0 ng/ml and 0.19 to 2.29 ng/ml in the last 10 days of gestation. The cause of this rise in transcortin is unknown [196]. It is interesting to speculate that it is brought about by increased cortisol secretion. This 10-fold increase in dialysable cortisol parallels the approximately 10-fold increase in the measurement of total glucocorticoid (Table 10.1). Fetal cortisol MCR remains unchanged over this period and therefore the cortisol turnover rate increases by a factor of 10. Maximum cortisol turnover rates which have been measured before normal parturition at term are 4.0 mg/24 h

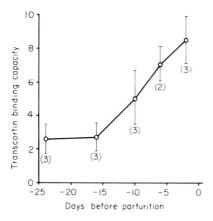

Fig. 10.4. Transcortin binding capacity (expressed in μg cortisol bound/100 ml plasma) of serial plasma samples from fetal lambs at various maturities. Figures in parentheses are numbers of fetuses. Means ± S.D. Reproduced with permission from Liggins et al. [391].

[391] and 9.2 mg/24 h [442]. The calculated turnover rates or BPR are directly proportional to the plasma cortisol concentration at the time of the experiment. Therefore there is a rapid increase in cortisol BPR on the last day of fetal life and values in individual animals can vary by a factor of 3 (see Tables 10.2 and 10.3).

As mentioned above, continuous infusion of 5.6 mg cortisol/24 h into fetuses of 110–130 days gestation does not produce delivery. Infusion of 11.2, 22.4 or 44.8 mg/24 h is, however, generally successful. Allowing for the fact that the systems on which cortisol is acting (see Chapter 12) may be unresponsive at this time, the data from calculation of fetal cortisol BPR at term and the doses of cortisol which must be infused to produce delivery are in good agreement.

10.2.7 Summary – four phases in the pattern of fetal plasma cortisol concentrations

Most investigators would accept the statement of Strott et al. [539] that 'Beginning 7–10 days pre-partum, fetal plasma cortisol levels progressively increased...'. Within the limitations of available methodology and with careful study of the data in the literature (papers referred to in Table 10.1 of cortisol values) it is possible to see suggestive evidence for the commencement of an increase in fetal plasma cortisol even a few days earlier than 10 days prepartum. In any event, it is possible to divide the pattern of observed fetal plasma cor-

Table 10.2.

Data summarised from Liggins et al. [391] for cortisol blood production rates (BPR) and transplacental transfer using isotope dilution techniques with continuous infusion of [^{3}H]cortisol.

Days before (parturition)	Plasma cortisol concentration (ng/ml)		Cortisol BPR (mg/24 h)		Transplacental transfer (mg/24 h)	
	Maternal	Fetal	Maternal	Fetal	Fetus → Ewe	Ewe → Fetus
−19 and −18	14.0	3.0	20.5	0.22	0.08	0.16
−17 and −7	6.9	5.2	11.3	0.88	0.11	0.21
−5 and −2	10.1	53.6	23.6	2.6	0.13	0.23
−1	9.8	75.8	—	6.2	—	0.07

Table 10.3.

Metabolic clearance rate (MCR) and cortisol production in the fetal and neonatal lamb.

Age (days before parturition)	No. of observations/ no. of animals	Plasma cortisol (μg/100 ml)	Mean wt. (kg)	MCR for plasma cortisol		Cortisol production/day (mg)	Cortisol production (mg/kg body weight/day)
				l/day	l/kg		
−4 or more	9/6	1.8 ± 0.3	2.2	58.1 ± 9.6	26.4	0.94 ± 0.15	0.43
−3	1/1	4.0	2.6	64.8	24.9	2.59	1.00
−2 and 1	4/4	12.2 ± 2.1	2.8	89.1 ± 32.9	31.8	9.17 ± 1.64	3.28
Newborn lambs:							
4–18 h	6/6	5.5 ± 0.6	2.8	78.5 ± 15.1	29.5 ± 4.8	4.3 ± 0.98	1.58 ± 0.34
15–17 days	5/5	1.5 ± 0.2	6.4	205.6 ± 49.7	30.8 ± 5.8	2.8 ± 0.55	0.41 ± 0.05
24–30 days	7/7	1.1 ± 0.1	9.1	202.8 ± 34.8	23.0 ± 3.9	2.1 ± 0.33	0.25 ± 0.05

Values given are means ± S.E.M.

tisol* changes into four phases, as was done in Chapter 9, Fig. 9.3. These are not demonstrated by all investigations. An initial phase of low plasma cortisol concentration between 120 and 135 days or so of gestation, phase 1; phase 2 is from 13 or so to 5 days before delivery; a third phase is a more rapid increase which occurs between 5 days before delivery and in the last 12-24 h of fetal life; lastly phase 4 represents a pronounced and very variable, final, stress-induced increment immediately before delivery.

Very little attention has been paid to the early minor changes in fetal plasma cortisol. This is unfortunate since the events responsible for the control of phase 2, the slow initial increment are of fundamental importance to the fetus. As discussed in Chapter 12, the fetal hypothalamo–hypophysial–adrenal axis plays a major role in the initiation of parturition in this species. The statement is based on several experimental observations. Interruption of the normal connection between the fetal hypothalamus and pituitary by inserting a silicone plate between them, or surgical ablation of the fetal pituitary or adrenal, all result in the prolongation of pregnancy with failure of parturition to commence. In contrast, premature delivery can be induced by infusing ACTH or cortisol into the fetus. It is therefore necessary to investigate the cause of both the growth of the adrenal and the commencement of increased cortisol secretion in phase 2.

10.3 Mechanism of action of glucocorticoids in the fetal sheep

In various species, glucocorticoids have been implicated in the following metabolic and developmental changes in the fetus and the neonatal period:

(a) Surfactant production from the Type II cells of the alveolus.
(b) Changes in the function of the thyroid axis.
(c) Initiating the endocrine changes in fetus and mother which are responsible for the delivery of the fetus.
(d) Conditioning of hypothalamic function.
(e) Methylation of noradrenalin to adrenalin.
(f) Causing involution of the thymus.
(g) Ionic changes in the amniotic and allantoic fluid.
(h) Development of gluconeogenic and other enzymes in the liver [343].

---

* In this section we are considering changes in physiologically active glucocorticoid. For simplicity this will be called cortisol. However, the remarks made about the various methodologies and possibilities of changes in plasma protein binding of cortisol should be recalled.

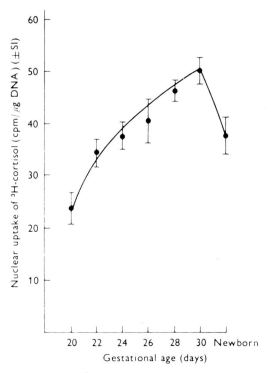

Fig. 10.5. Changes in uptake of [³H]cortisol (per unit DNA) by lung nuclei of the developing rabbit fetus. Each point is the mean ± S.E.M. of 5–6 determinations. For each determination tissue from lungs of all fetuses of the same litter was minced and the uptake of [³H]cortisol was measured. Further information on the methodology is reported by Giannopoulos et al. [242].

(j) Maturation of the small intestine.
(k) Changes in placental structure.
(l) Others [44, 48, 358].

In all these various procedures it is probable that glucocorticoids act by stimulating enzyme synthesis within their target tissue. The list of changes brought about by cortisol is intentionally completed by an all embracing category entitled 'others'. Since cortisol acts as the basic stimulus to the endocrine changes responsible for delivery, it is physiologically very useful to the fetus that the same agent should be responsible for inducing changes which will be vital for a free, independent existence after birth.

Activity of the glucocorticoid may be affected by the ability of the target tissue to bind the steroid. Fig. 10.5 demonstrates the doubling in uptake of

Glucocorticoids from the fetal adrenal gland 165

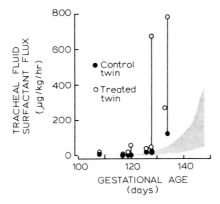

Fig. 10.6. Surfactant flux (μg/Kg/h) at different gestational ages in the fetal lamb. Surfactant flux is the product of tracheal fluid flow in ml/kg/h and surfactant concentration in μg surfactant lecithin/ml. The stippled area represents surfactant flux in untreated lambs. The effect of intra-abdominal infusion of dexamethasone (400 μg/day for 1.5–4 days) in twin pregnancies is shown. ○, treated twin, ●, control twin. Data reproduced with permission from Platzker et al. [474].

radioactive cortisol by lung minces from fetal rabbits of various gestational ages. There are several lines of evidence which show clearly that the development of surface active phosphoglycerides in the fetal lamb are dependent on the presence of cortisol. Hypophysectomised fetal lambs are deficient in pulmonary surfactant. In contrast, the infusion of ACTH or cortisol into pre-viable fetal lambs results in stimulation of pulmonary surfactant [386, 474]. Decapitation of the fetal rabbit in utero on day 24 of gestation leads to a decrease in osmiophilic inclusion bodies in the type II pneumocyte of the lung in the subsequent 5 days. These changes were not exactly paralleled by the changes in physical properties. ACTH reversed the histological changes [124]. Surfactant has been demonstrated in tracheal fluid of fetal lambs as early as 108 days gestation [474]. However, at this stage the quantities of surfactant are insufficient for the lamb to survive.

Lambs born after 140 days will generally survive [31]. At this time the final phases of plasma cortisol (3 and 4) have not yet commenced. It therefore seems that the phases of fetal cortisol which play a role in surfactant production are the low, basal concentrations of cortisol (phase 1 and possibly 2). Fig. 10.6 shows that the flux of surfactant into the tracheal fluid of fetal lambs (represented by the stipled area) increases steadily from about 120 days gestation. The rate of increase rises after 140 days gestation. There is considerable similarity between these changes and the superimposed fetal plasma corticosteroid concentrations. In addition, this figure shows that administration of the synthetic glucocorticoid dexamethasone directly to the fetus will increase surfactant flux up to 10-fold.

There is considerable evidence that glucocorticoids are involved with surfactant in the developing human fetus. Surfactant is a specific phosphoglyceride, dipalmitoyl lecithin, attached to a protein molecule [129]. The biosynthetic pathways have been studied best in the human. There are two major pathways for the synthesis of surface active lecithin [31]. In the early stages of gestation (22–34 weeks) the methyltransferase pathway is the major pathway. Around 33 weeks gestation, the choline phosphotransferase pathway becomes increasingly active, producing more and more of the specific dipalmitoyl lecithin. In Fig. 10.7, R represents a fatty acid residue. Whereas the choline phosphotransferase pathway produces a high yield of dipalmitoyl lecithin, the methyltransferase pathway incorporates diglycerides with fatty acids other than palmitate to produce lecithins which are less surface active than dipalmitoyl lecithin. Fig. 10.7 has been drawn in a way which shows the similarities of these two pathways. The starting point is ethanolamine in the methyltransferase pathway and choline in the choline phosphotransferase pathway. The structural difference between these two molecules is the three methyl groups on the nitrogen atom of choline. Different kinase enzymes produce phosphorylation of these substrates (1). The next step in each path is the same, catalysed by specific cytidyl transferase enzymes (2). Specific phosphotransferase enzymes (3) catalyse the production of phosphatidylethanolamine and phosphotidylcholine. The major difference at this step is that in the diglyceride utilised in the choline phosphotransferase (CPT) system is dipalmitate and the lecithin produced is the highly surface-active dipalmitoyl lecithin, whereas several different diglycerides are used in the methyltransferase pathway. The final step in the methyltransferase system (4) is the methylation of the nitrogen atom on the ethanolamine residue. This removes the difference between ethanolamine and choline at the beginning of the pathways.

The exact mechanisms whereby cortisol [72] and T_4 [73] stimulate surfactant production is not yet known. It is, however, likely to involve stimulation of enzymes active in the above pathways or in the synthesis of the protein moiety of surfactant.

Fig. 10.8 demonstrates that the administration of the synthetic steroid 9-fluoroprednisolone to fetal rabbits around 25 days gestation produces a 50% increase of incorporation of [^{14}C]choline into lecithin in the lungs of injected rabbits when compared with controls. CPT activity is also increased by over 30%. Enzyme activity is maximally increased in a little over 6 h and falls to control levels at 24 h. The time course of this activity is yet another indication

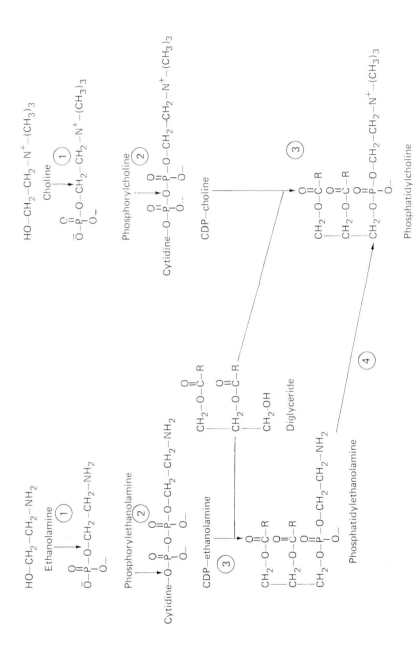

Fig. 10.7. Major steps in the two pathways responsible for the synthesis of lung lecithin. R represents a fatty acid residue. For a detailed description, see text.

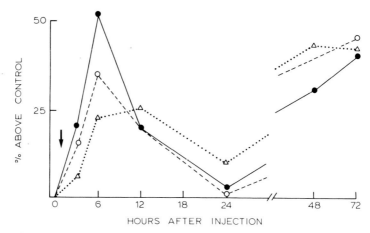

Fig. 10.8. Course of events in fetal rabbit lung after corticosterone injection. Fetal rabbits of ca. 25 days gestation were injected intramuscularly at t_0 with 9-fluoroprednisolone acetate. At various times thereafter, lung lecithin concentration ($\triangle \cdots \triangle$) and choline phosphotransferase (CPT) activity ($\bigcirc --- \bigcirc$) were measured and lecithin synthesis from [^{14}C]choline ($\bullet —— \bullet$) was estimated in lung slices. Mean data, from an average of 5 steroid-injected fetuses per point, are plotted as a percentage of the control value, as determined in saline-injected animals. Statistically significant points include CPT and [^{14}C]choline incorporation at 6 h and all parameters at 72 h. Reproduced, with permission, from Farrel [199].

of the rapidity with which the stimulation of enzyme activity by hormones can be noted when an end-point of activity which is early in the pathway is used. It used to be said that steroids required 48 h to stimulate enzyme induction. Fig. 10.8 shows that the effect can be demonstrated much ealier. However, when the end-point measured is the accumulation of lecithin within the lung, the latency is indeed about 48 h. A similar latency is required for the accumulated lecithin to be present in sufficient quantities to produce its physiological effect of lowering the surface tension of the fluid lining the lung alveoli.

Fig. 10.8 also shows that after the initial effect at 6 h, there is a second phase of enhancement of CPT activity 48–72 h after the injection. Several possible mechanisms have been put forward to explain this observation. For example, the injected steroid may enter the fetal amniotic fluid to be swallowed later and enter the fetal vasculature. Liggins and Howie have produced some evidence in the human to show that the effect of glucocorticoids on surfactant production is reversible. This observation is in keeping with the first peak of the effect shown in Fig. 10.8. A similar decline in the effect of corticosteroid induction of enzyme activity has been shown with tyrosine aminotransferase in the liver [299]. This decrease in activity is prevented in vivo by the continued production of steroid. In addition, certain vital systems, such as the glucocorticoid induction of parturition, involve self-perpetuating positive feedback loops which ensure that a particular system is brought to completion (see Chapter 12).

Administration of steroids to mother or fetus will affect surfactant production in several species including man. Failure of adequate surfactant production in newborn infants leads to the Respiratory Distress Syndrome (RDS). RDS is the cause of death in a large proportion of premature infants. The potential use of steroids to prevent RDS has been submitted to a clinical trial by Liggins and Howie with very encouraging results [393, 394]. Cortisol appears to increase the synthesis of surfactant but may work by other mechanisms as well. Therefore it is important to note that dexamethasone administration to pregnant rhesus monkeys has pronounced effects on various organs in the fetus [110] (see Chapter 12) and that administration of glucocorticoids to the newborn rat retards the growth of the brain and affects hypothalamic rhythmicity [371]. Administration of corticosteroid to pregnant rats results in the production of cleft palate in the fetuses [173]. However, use of dexamethasone in the latter half of human pregnancy probably occurs after most of the major structural differentiation has occurred. However, two points should be borne in mind. Firstly, differentiation of less visible structures, such as the brain, may still be continuing. Secondly, although the administration of natural estrogens in the early stages of pregnancy has not given rise to any reported side-effects, the administration of the synthetic estrogen stilbestrol has given rise to vaginal carcinoma in the female fetuses of treated women later in life [278, 281]. With reference to the first point of concern, it is reassuring that a follow-up study at $4\frac{1}{4}$ years showed no effect on I.Q. of the children of dexamethasone-treated women in premature labour to prevent RDS in these infants [394].

Parker and Noble [461] demonstrated that the administration of dexamethasone to pregnant rats from days 12 to 21 of gestation resulted in increased activity of the enzyme phosphoethanolamine-N-methyltransferase (PNMT) in the newborn pups. This is a further example of the ability of glucocorticoids to stimulate the activity of key enzymes. PNMT is responsible for the conversion of noradrenalin to adrenalin. PNMT activity was also increased in heart but not in brain. This is an interesting demonstration of differences in tissue response.

10.4 The adrenal cortex in the human and the sub-human primate fetus

10.4.1 Adrenal steroidogenesis

Both in vitro studies with primate adrenals and the estimation of steroids in umbilical blood show clearly that the fetal adrenal is capable of function at an early stage in gestation. In order that the fetal origin of a particular steroid can be proven it is necessary to demonstrate a positive arterio–venous difference across the umbilical circulation. Studies of infusion of labelled steroid precursors into the umbilical circulation of the previable human fetus have been reviewed by Solomon et al. [532]. Uptake and metabolism by the placenta and various fetal organs have been investigated. At this stage it is worth noting several points regarding these studies on the fetal adrenal in early gestation [532].

Firstly, activity of 3β-HSD and Δ^5-isomerase is low in the fetal adrenal at this time. Thus there is little conversion of the Δ^5-pathway steroids to Δ^4-steroids (see Fig. 10.1) in the fetal adrenal. Secondly, the placenta does possess these enzymes and can synthesise Δ^4-steroids including progesterone. This progesterone passes into both fetal and maternal blood. Progesterone is therefore available as a substrate for the fetal adrenal. Thirdly, although the Δ^5-pathway predominates within the fetal adrenal at this time, the fetal adrenal can produce cortisol [532]. The fetal primate adrenal has both 17-hydroxylase and 17-20-lyase activity and is therefore capable of producing androgenic C19 steroids. These steroids are aromatised in the placenta to produce estrogens (see Chapter 11).

Using chromatographic separation techniques together with competitive protein binding, it can be shown that cortisol is present in the human fetal adrenal at 10–18 weeks gestation [433]. In addition, a positive umbilical arterio–venous difference was demonstrated in plasma cortisol concentrations from fetuses at hysterotomy. In this study, the source of the plasma was elegantly confirmed by demonstrating a negative arterio–venous difference for progesterone. In addition, the negative arterio–venous difference for cortisone in fetal plasma confirmed the maternal origin of this steroid in fetal plasma as discussed below [430, 432].

The fetal primate adrenal cortex contains a hyperplastic zone which lies next to the medulla. This zone is termed the fetal or 'X' zone. It is generally contrasted to the area of cortex which will give rise to the zone glomerulosa, zona fasciculata and zona reticularis, which together are called the 'definitive cortex'. The fetal zone regresses rapidly after birth.

A recent study has demonstrated the importance of investigating both the zones of the fetal cortex described above [514]. When the definitive (or adult) and fetal zones are dissected out from 10–20-week-old human fetal adrenals there are, not surprisingly, some suggestions of differences between the two zones.

The in vitro output of the definitive zone was 5.7 ± 5.6 ng cortisol/100 mg/ 5 min in the initial period of incubation. This exceeded the cortisol output of the fetal zone but after 100 min the output of the definitive zone had fallen to less than the fetal zone. This latter observation again demonstrates the problems of studies undertaken with in vitro systems. It may well be explained on the basis that the definitive zone has a smaller store of precursors. The high standard error in the measurement of cortisol output demonstrates considerable individual variation, which may be due to the age of the material after abortion (0.5 to 11 h) and other problems associated with in vitro studies. It is of interest to compare the rate of cortisol production (5.7 ng/100 mg/5 min) with the output observed in the resting state during the study of sheep adrenals in vitro. Prior to the spurt in production around 140 days gestation, the output by the fetal sheep adrenal in vitro was 36 ng/100 mg/5 min. Addition of ACTH (25 mU/ml) to the incubates of human fetal definitive zone produced a 3–5-fold increase in cortisol production. ACTH had no effect on cortisol production by the fetal zone. This difference of in vitro ACTH responsiveness of the definitive and fetal zones may prove to be of considerable significance. The lack of ability of the fetal zone to respond to ACTH was accompanied by a lesser degree of ACTH binding by the cells of the fetal zone. This study again shows that the human fetal adrenal can produce cortisol even at 10–20 weeks gestation.

10.4.2 Plasma cortisol concentrations in human amniotic fluid and human umbilical plasma

There is now a large amount of data on glucocorticoid concentrations in amniotic fluid and also in the cord blood plasma of human infants delivering under various conditions. This data comes mainly from the laboratory of Dr. Beverley Murphy. Amniotic fluid cortisol concentrations showed a much better correlation with umbilical cord plasma cortisol concentrations than with maternal plasma cortisol [434]. Amniotic fluid cortisol concentrations showed a slow rise from 5 to 25 weeks gestational age (see Fig. 10.9). There followed a plateau until 37 weeks with a rapid rise from 38 to 40 weeks. The difference between amniotic fluid cortisol concentrations at 34–37 weeks and those measured after 38 weeks was highly significant.

These data demonstrate an interesting degree of similarity to plasma corticosteroid data from the chronically catheterised fetal sheep discussed in the

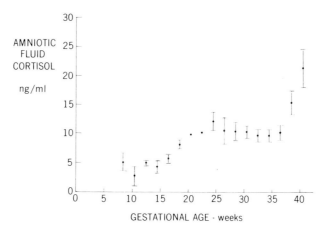

Fig. 10.9. Mean amniotic fluid cortisol at various gestational ages prior to the onset of labour. The bars show the S.E.M. Reproduced with permission from [434].

first part of this chapter. They agree in general with the only other extensive study on human umbilical plasma corticosteroid concentrations [515, 531]. In the first of these two papers the method of measurement involved competitive protein binding assay with solvent extraction of progesterone (one of the major factors which will cross-react as if it were cortisol) but no chromatographic separation. Plasma corticosteroids at term in umbilical arterial plasma in the group in spontaneous labour were 213 ng/ml. In the study in which Sephadex LH20 separation was used the concentration was 72.6 ng/ml. These values compare very well with data from Murphy et al. [430, 431]. Problems such as these should be borne in mind since other workers have failed to demonstrate significant differences in umbilical cord plasma concentrations in different situations [544].

Since the free, diffusible cortisol is the physiologically active fraction there is a need for data on free cortisol in the human. A comparison of this physiologically active fraction has been reported in cord plasma and maternal plasma in induced and spontaneous labour. There was no significant difference between the duration of the induced and spontaneous labours studied. Maternal physiologically active cortisol fraction was 132 ng/ml and 133 ng/ml in the spontaneous and induced labours, respectively. Umbilical arterial concentrations were 61 ng/ml (spontaneous) and 43 ng/ml (induced) [101].

This is a very important further indication that the fetus is actively increasing its own adrenal activity before the start of labour and that the increased cord plasma cortisol concentrations after vaginal delivery, as compared with caesarian section, do not simply reflect the stress of delivery. It is, however, necessary to bear in mind the possibility that the stresses provided by normal labour do not affect the fetus in the same way as those resulting from labour induced by oxytocin or prostaglandin. In the experimental sheep model, part of the terminal increment in fetal plasma cortisol concentrations (Phase 4 in Fig. 9.3) is due to secondary effects on the hypothalamo–hypophysial–adrenal axis.

Even though it appears that plasma cortisol concentrations follow a similar, though obviously not identical, pattern in the human to that documented in the sheep fetus, it should not be too readily concluded that the actions of cortisol are the same in the two species. This question will be reviewed in relation to parturition in Chapter 12. At this point it is appropriate to discuss one major difference between the primate and the sheep – namely placental permeability to glucocorticoids.

10.4.3 *Considerations of placental transfer of cortisol*

The fetal sheep is protected against the effects of large changes in maternal cortisol by the diffusion resistance offered by the placenta. In the human, a different mechanism exists. Cortisone concentrations are greater in umbilical vein plasma than in umbilical arterial plasma in neonatal humans. This observation suggests the presence of an active 11β-dehydrogenase in the human placenta. This enzyme converts cortisol to cortisone by converting the hydroxyl group at position 11 to a ketone group (see Fig. 10.1). The involvement of this enzyme during placental transfer of cortisol has been confirmed by the administration of labelled cortisol to pregnant humans immediately before delivery [432]. It was demonstrated that 85% of the cortisol crossing the placenta was converted to cortisone. Since cortisone is inactive as a glucocorticoid and the fetus cannot reverse this oxidation step, the fetus is, to a large extent, protected against any changes in maternal plasma cortisol concentrations [60, 106, 420, 432, 460].

10.4.4 *Pituitary control of the fetal adrenal*

ACTH has been demonstrated in plasma of human fetuses removed from the uterus by hysterotomy as early as the end of the first trimester [590]. Concentrations were about 250 pg/ml at this time, whereas at term the concentrations were only half this value, about 120 pg/ml (see Fig. 10.10). Fig. 10.10 also demonstrates the growth of the human fetal adrenal gland. It should be borne in mind that the observation of very high cord blood ACTH concentrations may reflect the effects of hysterotomy and the subsequent time spent in the extrauterine environment before blood was drawn, rather than reflecting the undisturbed state in utero. The same difficulty lies in comparing maternal and neonatal cord blood concentrations in an attempt to assess any possibility of placental transfer.

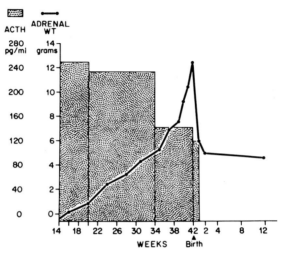

Fig. 10.10. Relationship of fetal and neonatal plasma ACTH concentrations to adrenal weight in the human. Reproduced with permission from [590].

As mentioned above, ACTH will stimulate cortisol production from definitive cortex in vitro but not the fetal cortex. The fetal cortex regresses rapidly after birth, suggesting that some trophic factor which had been operating in utero is now missing. It is therefore possible that a placental factor plays a role in the maintenance of the primate fetal zone. Such a placental factor may depend on fetal pituitary control. It is known that the fetal zone regresses prematurely in anencephalics, an observation which at least suggests fetal involvement [122, 300].

In summary, much emphasis has been placed on the structural difference between the primate and ruminant fetal adrenal with respect to fetal zone. Whether this structural difference reflects a significant difference in terms of the production of cortisol needs further investigation. With regard to the control of both the fetal and definitive cortex, there are very few controlled experimental data in the primate. The observations available from hypophysectomised fetal monkeys and human anencephalics can be explained by several different mechanisms, most of which bear close resemblance to the ideas considered in relation to the control of the fetal lamb adrenal. The primate adrenal growth curve (Fig. 10.10) shows similarities to that of the lamb. It is very likely that several trophic factors will be involved in its control. Whether cortisol from the fetal primate adrenal plays the same role in the initiation of parturition, as it is thought to do in the sheep, will be discussed in Chapter 12.

Fetal estrogens, progesterone and prostaglandins in the sheep and primate

The two previous chapters have dealt with the development of the adrenal axis in the fetal sheep and in the fetal primate. Special attention has been paid to those changes in the later stages of gestation that are thought to be intimately concerned with the initiation of parturition. Involvement of the hypothalamo–pituitary–adrenal axis in the processes which are supportive once parturition has begun was also considered. In this chapter three further factors present on the fetal side of the placenta will be discussed. They are estrogens, progesterone and prostaglandins. The first two are definitely related to fetal adrenal function and the third may yet prove to be, although there is no evidence to support this suggestion at the present time. In addition, it is necessary to consider to what extent all three factors may play a role in the processes of parturition. These factors will be discussed first in the sheep and then in the primate.

11.1 Sheep

11.1.1 Estrogens in the fetal sheep

Maternal plasma estrogen concentrations in the sheep are low until the last 1–2 days of gestation [107, 391, 456]. This situation is very different from the primate. Great care should be exercised in extrapolating the mechanisms of estrogen production in the ovine pregnancy to those which occur in human pregnancy. A similar cautionary note was mentioned with relation to the metabolism of cortisol. In addition to the differences in the two individual systems in the sheep and primate, it has now been demonstrated that glucocorticoids play a vital role in the initiation of changes in estrogen levels which precede delivery in the sheep. No such effect has yet been definitely demonstrated in the primate.

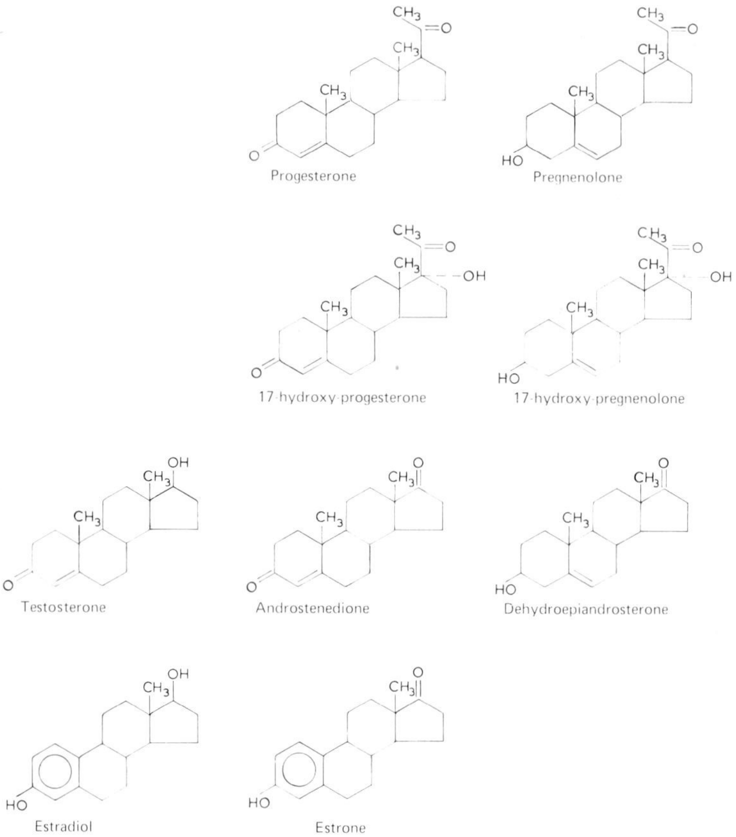

Fig. 11.1. Structure of the major steroids involved in the biosynthetic pathway of progesterone to estrogens.

Figs. 11.1 and 11.2 show the important metabolic pathways that have to be considered in relation to the production of estrogens in the sheep fetus [22, 329]. In this species during the latter stages of pregnancy, circulating maternal estrogens are of fetal or placental origin. Fig. 11.3 demonstrates that, at normal parturition, fetal plasma estrogen concentrations rise before maternal plasma estrogens. In addition, administration of glucocorticoid to adrenalectomised sheep fetuses results in the initiation of delivery with an increase in fetal plasma estrogens which again precedes the rise in maternal estrogens [218].

Glucocorticoids administered to the intact fetal sheep result in increased activity of the C17-hydroxylase and C17,20-lyase enzymes in the placenta [21, 22]. It therefore appears that glucocorticoids will increase the placental production of C19 estrogen precursors. Fetal plasma concentrations of one of

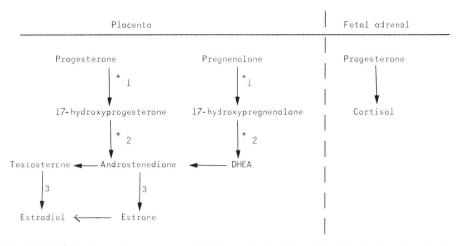

Fig. 11.2. This figure shows the possible interrelationship of the fetal adrenal and placenta in the production of estrogens from progesterone in the sheep. These ideas are based on the work of Turnbull, Anderson, Flint and their colleagues at Oxford. Major stages in steroidogenesis in the fetal sheep adrenal and placenta which are responsible for the production of estrogens are shown. *Enzyme steps upon which cortisol is considered to act [21, 22, 219]. (1) C21 steroid 17α-hydroxylase; (2) 17,20-lyase; (3) aromatizing enzymes – no effect of cortisol has yet been shown on these enzymes.

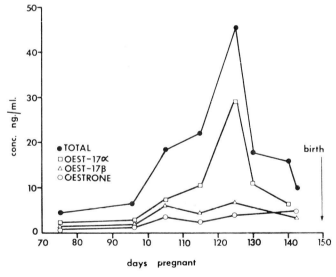

Fig. 11.3. Concentrations of fetal plasma sulpho-conjugated and sulpho-glucosiduronates of estrogens at different gestational ages. Reproduced with permission from Findlay and Seamark [204].

these, androstenedione, rise from about 100–500 to 1000–2000 pg/ml before spontaneous parturition or ACTH-induced delivery [388].

It is possible that the investigation of changes in plasma concentrations of precursors or intermediates in this pathway will not be a very rewarding approach to the problem. Placental biosynthesis of progesterone probably provides enough precursor for the system and the intermediates may not be released into the blood in significant amounts. The evidence cited above suggests that glucocorticoid enhances the enzymatic conversion of progesterone to androstenedione. Although it has been known for a long time that the sheep placenta contains the necessary aromatases to convert C19 estrogen precursors to estrogens [204] it is as yet uncertain whether glucocorticoids will increase the activity of placental aromatases which constitute the final step in the synthesis of estrogens (see Fig. 11.2). It would be very useful to know which of these enzyme steps is rate-limiting in the production of estrogens. In addition to their actions on various enzyme systems, estrogens appear to influence fetal growth [2], and uterine blood flow [250, 479, 491, 502]. This action on uterine blood flow may help to maintain placental blood flow during labour [556].

Findlay and Seamark [204] have investigated the fate of radioactively labelled androgens infused into a fetus at 114 days gestation. The radioactivity was incorporated predominantly into estradiol-17α, estradiol-17β and estrone. They also observed rapid sulphation of these estrogens as well as a greater conversion to estrogens of androstenedione than dehydroepiandrosterone (DHEA). Wong et al. [594] have shown that the overall conversion rate of labelled androstenedione infused into the fetus to estrogens in fetal plasma is very low (not greater than 1%). This may be an indication that under physiological conditions, the C19 precursors are generated within the placental tissues themselves. An alternative suggestion is that the physiological C19 precursor is a steroid other than androstenedione, testosterone or DHEA. One study has demonstrated a fall in fetal plasma testosterone from 4 to 1 ng/ml in the last two days of gestation. In contrast, no fall in plasma testosterone occurred when ACTH (4–5 μg/h) was infused into the fetus for 5 days [539].

Estrogens exist in plasma in a free form, bound to plasma proteins and as conjugated estrogens. The major estrogen in fetal sheep plasma is probably estradiol-17α [203]. The major form of conjugate is the sulphate (Fig. 11.4). The placenta contains both sulphokinase and sulphatase enzymes (Fig. 11.4), whose activity maintains the balance in the various fractions in which estrogen circulates in the plasma. As with other hormones which are conjugated, it is considered that the conjugated form of the steroid is inactive and has to be hydrolysed to the unconjugated form to retain metabolic potency.

Both sulpho-conjugated and unconjugated estrogens in fetal plasma rise in

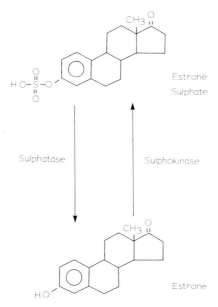

Fig. 11.4. Interrelationship of estrone and estrone sulphate. Similar conversions exist for the other estrogens.

the last 36 hours or so of gestation [56, 151, 483]. Conjugated estrogen concentrations may rise to 50 ng/ml whereas unconjugated estrogens may reach only 1% of this concentration, around 500 pg/ml. Similar increases occur during dexamethasone-induced delivery [151]. Whereas labelled estrogens pass readily from fetal to maternal circulation, there does not appear to be much passage in the opposite direction [204]. The passage of estrogens from fetus to mother may account for the fall in fetal plasma estrogen conjugates which follows the peak concentrations shown in Fig. 11.3. It should be noted that the highest concentrations of fetal estrogens demonstrated in this study occur well before the fetal plasma cortisol increase in the last five or so days of gestation (Fig. 9.3). There is a need for a careful relationship of fetal plasma and tissue estrogen concentrations in a sequential study of individual animals undergoing normal delivery. Some data is available which suggests that in chronic studies the increase in fetal unconjugated total estrogens only occurs in the last 3–4 days of gestation [483].

The absolute concentrations of estrogen reached in the maternal circulation vary considerably between animals. Similarly, individual maternal plasma progesterone concentrations studied show marked variation, even for single pregnancies. One observation of potential importance is that the induction of delivery following ACTH administration to fetuses of 112 to 124 days gestation induced labour but free estradiol-17β concentrations remained below the limit of sensitivity of the assay. In contrast, in 9 out of 10 ewes at 130 days gestation or more, fetal plasma estrogen concentrations rose following infusion of dexamethasone into the fetus [391a]. It may, however, be concluded that a rise in fetal plasma estrogens is not a necessary prerequisite to induced delivery at less than 130 days. This does not however rule out the possibility that estrogen concentrations increase within the placenta and have a local effect on the myometrium either directly or through the production of prostaglandins (see section 11.1.3).

In one animal in which parturition was blocked by the administration of progesterone to the mother, maternal plasma estrogen concentrations rose to three times those normally found at delivery [391a]. It is interesting to speculate on the possibility that this was due to the greater amounts of progesterone available as substrate following the administration of exogenous progesterone. Alternatively, the greater production of estrogen may simply reflect the longer time interval available for estrogen biosynthesis when labour was blocked with progesterone.

In summary, fetal plasma estrogen concentrations, both conjugated and unconjugated fractions, rise in the last 36 hours of gestation, probably as a result of conversion of progesterone to estrone and estradiol in the placenta. This rise in plasma concentration is controlled by fetal glucocorticoids which induce a level of increased activity in the placenta of 17-hydroxylase and 17,20-lyase enzymes and possibly of aromatase enzymes as well. Sulphatase and sulphokinase activity maintain a balance of conjugated and unconjugated estrogen in fetal plasma. Estrogen synthesised in the placenta can traverse the placenta and appear in the maternal circulation. In contrast, estrogen does not readily pass from the ewe to the fetus. For this and other reasons it is considered that the feto–placental unit is the major source of estrogens in the sheep.

11.1.2 Progesterone in the fetal sheep

As mentioned previously, the placenta is the major site of progesterone production in the later stages of gestation. Removal of the maternal ovary after the 60th day of gestation does not precipitate delivery [168] as it would do in species such as the goat or rabbit, in which the corpus luteum is required throughout pregnancy [147]. In this respect the ewe and the human are similar. Certain species occupy a somewhat intermediate position, such as the cow. The cow is an interesting species with respect to feto–maternal metabolism of progesterone and estrogen. The physiology of parturition in this species has been discussed in detail elsewhere [440, 573].

Plasma progesterone concentrations in the fetal sheep are low. They rarely exceed 1 ng/ml [539]. This is due to the activity of a 20α-hydroxysteroid dehydrogenase (20α-HSD) which is present in the fetal red cells and in other tissues [437, 469]. The presence of this enzyme, in maternal as well as fetal tissues, complicates calculations of clearance rates and blood production rates for progesterone in both the fetal and maternal compartments. Detailed studies on the kinetics of progesterone metabolism in the pregnant sheep have demonstrated that the production rate of progesterone in the maternal compartment increases nearly 5-fold between 7 and 145 days gestation [57, 58]. Unfortunately there are no detailed studies of the kinetics of progesterone production in the fetal compartment.

The conversion of placental progesterone to estrogen has been discussed above. The products of the 17α-hydroxylase step in this conversion have been demonstrated in both fetal and maternal plasma. The concentration in fetal plasma generally exceeds the concentration in maternal plasma during both normal and dexamethasone-induced delivery. It has been calculated that this pathway accounts for a mean of 49.1% (range 19–98%) of the progesterone disappearing from the maternal circulation [219]. It is therefore possible that the glucocorticoid-stimulated production of estrogen by the ovine placenta occurs by mechanisms which also produce decreased output of progesterone from the placenta and a subsequent fall in maternal plasma progesterone concentrations. The time relations of the fall in maternal progesterone concentrations, which occurs before the rise in maternal estrogen, would support this possibility. The variability in the range of the calculated percentage of progesterone utilised by this mechanism may account for the great variability in the fall in maternal progesterone which accompanies normal delivery in the sheep. Delivery may occur whilst there is still 3–4 ng/ml progesterone in maternal plasma.

In premature induction of parturition following the administration of dexamethasone or ACTH to the fetus, maternal plasma estrogen concentration may not rise (as in the animals of 124 days gestation or less, mentioned above). Maternal plasma progesterone concentrations do, however, fall. This dissociation in this instance of the fall in maternal plasma progesterone concentration from the subsequent rise in maternal plasma estrogens could be explained if all the three enzymes in the pathway from progesterone to estrogen (Fig. 11.2) were not working effectively before 130 days gestation. Plasma progesterone might then fall, due to induction of efficient activity of 17α-hydroxylase in the presence of inefficient action of either or both 17,20-lyase and aromatase.

11.1.3 Prostaglandins in the fetal sheep

The structure of this important group of highly active molecules is given in Fig. 12.7. They play many important roles, especially in reproductive physiology [192, 362]. Infusion of prostaglandin $F_{2\alpha}$ ($PGF_{2\alpha}$) into chronically catheterised fetal lambs will produce uterine contraction (Fig. 11.5). $PGF_{2\alpha}$ probably plays an important oxytocic role in the sheep (see Chapter 12). Although prostaglandins are ubiquitous in mammalian tissues [353] very little is known regarding their production and function within the fetal compartment. Liggins et al. [391] have shown that PGF is present in the fetal layers of the placental cotyledon and the concentration increases during the induction of delivery by infusion of dexamethasone into the fetus. At the present time it is not known whether PGF is synthesised in the fetal layers or whether it has diffused or been transported there after synthesis in the maternal layers of the cotyledon where the concentrations are greater than in the fetal layers [391]. The stimulus to the synthesis of PG is probably estrogen, produced to the administered glucocorticoid. The mechanism of production of PG by the decidua is discussed in greater detail in relation to the primate fetus.

In the adult mammal, PGs are metabolised by many different tissues. A prominent, though not the sole, site of metabolism is the lung. Enhanced in-

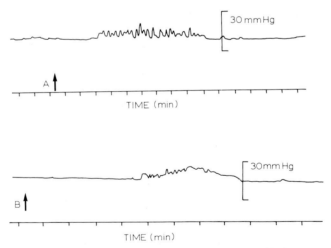

Fig. 11.5. Record of uterine activity in a pregnant ewe at 141 days gestation. Fetal vascular catheterisation was performed at 135 days gestation. (A) 20 μg $PGF_{2\alpha}$ was injected into the fetal inferior vena cava; (B) shows the effect of 100 mU of oxytocin administered to the fetus on the same day. This effect is shown here to compare the oxytocic effects of administration of oxytocin and $PGF_{2\alpha}$ to the fetus (see Chapter 7).

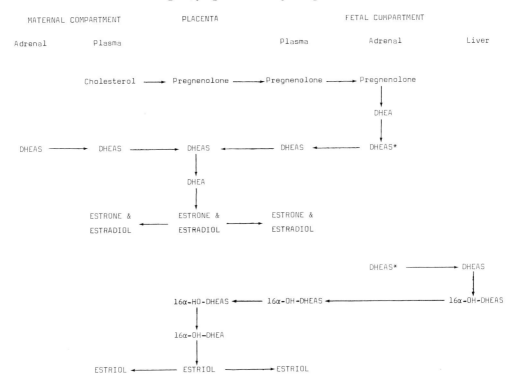

Fig. 11.6. Diagrammatic representation of the tissues in which the major steps in the production of estrone, estradiol and estriol occur in the primate fetus. (A) Production of estrone and estradiol takes place in the placenta using DHEA which is produced in both the maternal and fetal adrenals. (B) *Estriol production in the primate fetus. The starting point in this flow diagram is DHEAS which is produced as shown in (A)*. Placental estriol production is totally dependent on the provision of 16α-OH-DHEA from the fetal liver. This dependence on fetal liver function demonstrates why measurements of maternal plasma estriol are a better index of fetal well being than measurements of maternal plasma estradiol and estrone concentrations.

activation of PG may occur during pregnancy [59]. In the fetus, pulmonary blood flow is very low, which may be a factor in maintaining high peripheral PG concentrations at the time of delivery. A major site of PG metabolism in the sheep is the myometrium. Compared with the primate, very little PG is metabolised by the fetal membranes in the sheep (see below). This difference may account for variation in the response of the sheep and the primate to maternally administered PG. In addition, lower metabolism of PG by the fetal membranes in the sheep when compared with maternal tissue [354b] may assist the passage of PG from fetus to mother, thus enabling it to exert an oxytocic effect. PGs also probably play a function in the control of the fetal circulation. For example, it appears that PG plays a role in the closure of the ductus arteriosus after birth [191, 454, 533].

11.2 Human and sub-human private

11.2.1 Production of estrogens by the feto–placental unit in primates

Table 11.1 sets out the basic differences between the production of estrogens within the feto–placental unit of the sheep and the primate. The enzymatic conversions which have to be carried out are the same as in the sheep (see Figs. 11.2 and 11.6) but there are two main differences. Firstly, the tissue in which the individual conversion steps occur may differ. Secondly, the fetal primate liver has a potent 16α-hydroxylating system which is not active in the sheep.

As mentioned before, the fetal sheep adrenal possesses 3β-HSD and therefore can synthesise progesterone from pregnenolone (Fig. 10.1). In the case of the primate fetus there are two possible adrenal C21 precursors, pregnenolone produced within the gland itself and progesterone obtained from the fetal plasma in which it is present at high concentrations (1 µg/ml in the human at term; see below). The fetal adrenal possesses both 17-hydroxylase and 17,20-lyase activity and therefore the primate adrenal can synthesise C19 steroids [455, 523, 532]. The major steroid product from pregnenolone is dehydroepiandrosterone (DHEA) and its sulpho-conjugate, DHEAS. The placenta lacks these two enzymes and cannot produce C19 steroids. However, placental aromatases are responsible for the final transformation of C19 steroids to estrone and estradiol [523, 171].

Table 11.1.
Comparison of estrogen production in the feto–placental unit of the sheep and the primate.

	Sheep	Primate
Major estrogens	Estrone	Estrone
	Estradiol	Estradiol
		Estriol
Site of 17-hydroxylation	Placenta	Fetal adrenal
		Maternal adrenal
Site of 17,20-lyase activity	Placenta	Fetal adrenal
		Maternal adrenal
Site of aromatase activity	Placenta	Placenta
Site of 16α-hydroxylation	—	Fetal liver

Sequential studies of the production of estrogens by the primate fetus at different ages of gestation have not yet proved possible. Cord blood plasma testosterone concentrations are higher in the male than the female fetus of the rhesus monkey [489]. Plasma concentrations increase between 120 days and term. These observations suggest that the majority of testosterone in the male is derived from the testis (see Chapter 3). The demonstration of an increase in plasma testosterone in both sexes suggests that adrenal C19 precursors may rise as term approaches. However, testosterone is probably not a very important C19 estrogen precursor.

11.2.2 Relationship of fetal estrogens to changes in maternal plasma estrogen

C19 estrogen precursors are also supplied to the placenta by the maternal adrenal. For this reason, measurements of estradiol or estrone in maternal plasma are not a good indication of fetal function. However, 16α-hydroxylation appears to be confined to the fetal liver. The product of 16α-hydroxylation of DHEA is 16α-OH-DHEA. This is then converted in the placenta to estriol. For this reason, measurements of estriol in maternal plasma or urine are a good indication of feto–placental function (Fig. 11.6).

One study of cord blood plasma estrogens demonstrated that unlike the sheep, the maternal to fetal concentration ratio for plasma estrogens was 10 : 1 in the rhesus monkey. A slow increase in fetal estradiol was demonstrated over the last 60 days of gestation. In general, umbilical vein plasma estrogen concentration exceeded umbilical arterial concentrations, again demonstrating the placental origin of fetal estrogens [490]. It was mentioned above that studies in the sheep have demonstrated a greater passage of estrogens across the placenta in the fetal to maternal direction than in the reverse direction [204]. Knowledge of the permeability and the rates of transport of estrogens across the primate placenta would be of use in interpreting the small changes in fetal estrogens demonstrated.*

Measurements of maternal plasma estrogen concentrations throughout gestation in the rhesus monkey have shown that there is a gradual rise in the latter part of pregnancy with a more pronounced increase in the last 10 days [84]. If, as suggested above, the cause of this rise is an increased production of C19 steroids from the adrenal, the signal which initiates the increased adrenal activity is as fundamental a question in relation to the delivery of primates as it was in the sheep (see Chapter 10).

---

* Placental sulphatase enzymes play an important role in the production of maternal estrogens from sulpho-conjugates in fetal plasma [226].

In the human there has been much debate as to whether there is an increase in maternal plasma estrogen concentration in the days preceding delivery. In one study maternal plasma estradiol concentrations have been shown to increase over the five weeks which precede normal labour at term [559]. The same group of workers have shown a rise in maternal plasma estrogens in women in premature labour [545]. Various factors complicate attempts to determine whether there is an increase in maternal estrogens in the human at term. From the data obtained with much more frequent sampling in sheep, it is clear that maternal plasma estrogen concentrations can vary considerably within the same animal even during the rapid increase that occurs over the last hours of gestation. In addition, a large proportion of the estrogen in the maternal circulation is bound to sex-hormone-binding globulins. This influences the interpretation of transplacental concentration gradients. Finally, comparison of results between different groups is difficult due to the different specificities of the antisera used and the details of extraction and steroid separation employed.

11.2.3 Progesterone in the fetal primate

Human term cord blood plasma concentrations of progesterone are high – more than 1 µg/ml. This progesterone is produced by the placenta since there is a negative arterio–venous difference when concentrations in the umbilical artery and vein are compared [261]. There is an interesting sex difference in the human. Although umbilical vein plasma progesterone concentrations are the same (1.5 µg/ml) in male and female infants, the concentration in the umbilical artery of female infants is less (0.7 µg/ml) than in the male (0.9 µg/ml). The significance of this observation is unknown. It may either reflect increased metabolism by the female fetus prior to or during delivery or possibly increased secretion by some tissue in the male. Maternal plasma progesterone concentrations are 100–150 ng/ml just before delivery.

In the fetal rhesus monkey from 110–163 days of gestation, plasma progesterone concentrations were also higher in female umbilical arterial plasma than in the male (11.1 vs. 5.2 ng/ml). These values are about 1% of the absolute concentrations observed in the human. In contrast to the human, a sex difference also exists in umbilical vein plasma in which the progesterone concentrations are 22.0 ng/ml in the female and 7.3 ng/ml in the male. Maternal peripheral plasma progesterone concentrations showed no difference with respect to the sex of the fetus, and ranged from 1.0–5.7 ng/ml.

As was the case with observations of maternal estrogen concentrations, there is much conflicting data on the question of whether there is a fall in maternal progesterone concentration before delivery in the primate [149]. There is

CHAPTER 12

Parturition and the feto–placental unit

12.1 Introduction

The myometrial cell is the final effector cell in the processes which result in the efficient completion of parturition. The final stages of delivery in several species are accompanied by expulsive abdominal effort (Fig. 12.1). However, in the sheep, this terminal phase has been preceded by 6–12 hours of gradually increasing uterine activity [284, 286, 287]. Most of the major factors, both fetal and maternal, which affect the function of these contractile cells have been dealt with individually in the preceding chapters, especially Chapters 7, 9, 10 and 11. A degree of overlap with the material in these chapters is therefore inevitable if the composite picture is to be obtained.

This chapter will explore the following question 'What is the timing mechanism, or clock, that decides when the fetus will commence its journey down the birth canal; a journey which results in its appearance as a free-living organism in the external world; a journey potentially more hazardous than any other that this individual will take during its life'. In utero the conditions have suited the fetus well – unless some pathological process has been present. Although it will be difficult to prove the point – as with the proof of all negative conclusions – it does not seem at the present time that the fetus is becoming short of any particular metabolite(s) immediately prior to delivery. It is difficult to exclude a shortage of availability of a particular metabolite but the current evidence suggests a mechanism which has more biological significance. Certainly, if parturition is delayed for any reason, the fetus may eventually find itself short of nutrients. However, it now appears that the fetus plays a far more active role in determining the timing of the onset of uterine contractions. From a teleological viewpoint, this is a far more 'sensible' system for evolution to adopt since it allows the possibility that parturition is linked to

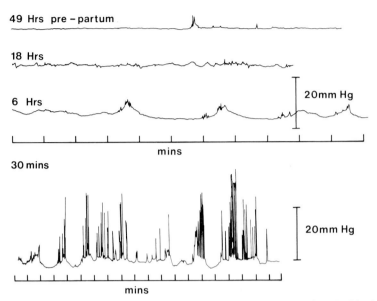

Fig. 12.1. Development of uterine contractility during fetal infusion of cortisol in the ewe. Very little uterine activity exists 48 h before delivery. In normal parturition contractions do not commence until about 6 h before delivery. In the last 30 min the uterine contractions were much more frequent, occurring every 1 to 2 min with superimposed abdominal efforts.

the maturation of physiological systems which will be required for an independent extrauterine existence.

This chapter will follow the general pattern observed throughout the book. Initially we shall discuss the fetal sheep. In no field has the chronically catheterised fetal sheep preparation yielded more information of fundamental physiological importance than in the consideration of the mechanisms involved in parturition. No experimental preparation is better suited to the administration of active and inhibitory agents at physiological concentrations, or to monitoring the changes in fetal and maternal endocrinology, together with detailed observation of myometrial activity. Some information obtained from polytocous species, especially the rat and rabbit will also be included. This will be followed by a section describing experiments in which removal of the fetus has been undertaken. For obvious reasons, data from such fetectomy experiments are often invoked when evidence for fetal initiation of the events leading up to parturition is proposed. Finally, an assessment will be made as to the extent to which these observations in experimental animals, particularly those obtained from the sheep, can be applied to work which has been conducted in the primate and in particular in the human.

12.2 The initiation, maintenance and successful completion of parturition in the sheep

12.2.1 Historical background

Malpas suggested in 1933 [406] that anencephaly in the human is associated with post-maturity. It has been known for some time that certain genetic abnormalities in Holstein–Friesian and Guernsey cattle are also associated with prolonged gestation [297, 358, 387]. These developmental anomalies affect the fetal adenohypophysis. However, the most spectacular experiment conducted by nature is the result of the teratogenic effect of an alkaloid from the vetch, *Veratrum californicum* [67–69]. If this grass is eaten by pregnant ewes during a critical period before the sixteenth day of pregnancy, gross fetal malformations occur. If it is ingested later in pregnancy the alkaloid has no apparent effect. There is a susceptible period of embryonic development at which the alkaloid produces abnormal development of the brain. The hypothalamic–pituitary region is one of the most badly affected areas in a fetus which is a grossly deformed caricature of a lamb. Other abnormalities can be noted, especially single central eye (cyclopia) and a massive tongue (macroglossia) [67]. In the presence of these malformations, parturition does not occur and the fetus must be delivered by caesarian section. Normal gestation length in the sheep is usually around 147–150 days, with slight breed differences. Cyclopian fetuses have been delivered by caesarian section as late as 250 days gestation.

These malformed fetuses have a dislocated pituitary. As with the human anencephalic fetus they will certainly constitute a very heterogenous group, although no hormonal measurements have been made on either the fetus or the mother. It appears that at least a portion of the fetal pituitary is present and these fetuses are thus more similar to stalk-sectioned experimental fetuses than hypophysectomised fetuses. The fetuses continue to grow in utero and it would be of great interest to obtain information regarding fetal adrenal growth and secretion. Other abnormalities during pregnancy have been reported from the African Continent, resulting from the ingestion of alkaloids at later stages of pregnancy [50].

Clues such as these led to the exploitation of the fetal sheep preparation [386, 392]. The results of approaches such as fetal hypophysectomy [395, 396], fetal pituitary stalk section [391] and fetal adrenalectomy [174], together with the effect of infusion of glucocorticoids into the fetus have been discussed in Chapters 9, 10 and 11, in relation to the development of fetal adrenal function (Fig. 12.2).* Those aspects directly related to the processes of parturition will now be reviewed.

---

* Hypophysectomy of the pregnant ewe does not affect the length of gestation [169].

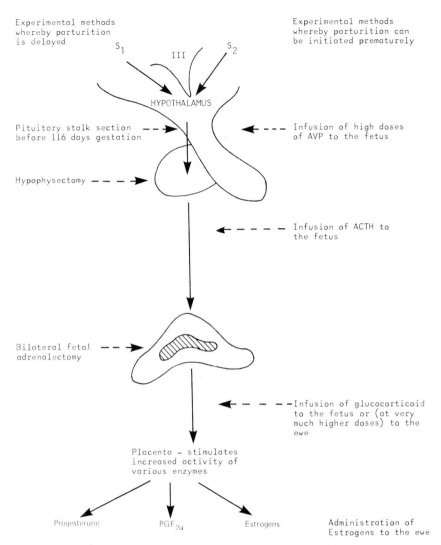

Fig. 12.2. The various levels of the fetal hypothalamo–hypophysial–adrenal–placental axis at which important events may occur prior to parturition. On the left are shown experimental techniques whereby parturition can be delayed. On the right are methods of initiating parturition prematurely. S_1 and S_2 represent the various stimuli which may act on the hypothalamus. III is the third ventricle.

12.2.2 The central role of the fetal adrenal

Based on the evidence discussed earlier we may assign a central role in the processes of parturition in the sheep to the fetal lamb adrenal. In discussing these processes it will help if we consider the simplified flow-diagram shown in Fig. 12.3. The INPUT side of this figure is the aspect concerned with the question

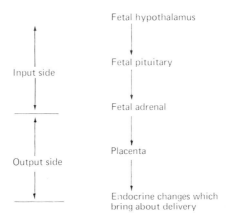

Fig. 12.3. Simplified flow diagram in which the fetal lamb adrenal is assigned a central role in the processes involved in the initiation of parturition. We may consider the events which bring about the increased secretion of cortisol from the fetal adrenal as the INPUT side; the mechanisms whereby the increased secretion of cortisol produces the endocrine changes which bring about delivery may be considered the OUTPUT side.

of the timing of the onset of delivery which was posed at the beginning of this chapter. The OUTPUT side is concerned with the various hormonal and metabolic regulators which affect the contraction of the myometrial cell. Control of the contractile property of this cell is carried out by stimulatory and inhibitory factors [509]. Most of the important factors occur in all species. There is, however, a bewildering array of different patterns of plasma hormone concentrations; of the changes that take place at delivery in the plasma hormone concentrations and their blood production rates. Despite the magnitude of the problem, a result of the several factors involved, and the difficulty of correlating in vitro and in vivo studies, progress is being made in the formulation of general principles describing the electrophysiology and biochemistry of the myometrial cell. This is a specialised form of smooth muscle cell and the biophysical techniques used for the investigation of other forms of smooth muscle are now being applied to it [348]. Several of the major factors, both fetal and maternal, which affect the myometrial cell have been dealt with individually in the preceding chapters (especially Chapter 11). No attention has been paid to various pharmacological agents, many of which have a long history [152].

12.2.3 The input to increased fetal adrenal secretion of glucocorticoids

Hypophysectomy of the fetal lamb before 142 days gestation will result in adrenal hypoplasia and prevents the occurrence of parturition at term [396]. Fetal plasma ACTH concentrations fall to less than 0.34 pg/ml after hypophysectomy. In addition, pituitary stalk-section without pituitary infarction delays parturition [391]. No data is available at the present time of ACTH concentrations in the pituitary stalk–sectioned fetus but the continued growth of the fetal adrenal suggests the continuation of at least part of the normal trophic control to the adrenal in this situation.

In Chapter 10, a detailed consideration was undertaken of the possible factors which may control the development of adrenal sensitivity to ACTH and bring about the increased fetal secretion of cortisol. It was stated that, at the present time, there is no evidence that increased fetal pituitary secretion of ACTH is responsible for the rise of fetal adrenal cortisol output. The observation that ACTH replacement to the hypophysectomised fetus initiates parturition is of considerable importance [330]. However, three points are worth noting. Firstly, fetal pituitary ACTH secretion rates are probably around 1 µg/h rather than the 10 µg/h used in this study. Secondly, parturition was not normal in these three animals. Thirdly, as is always the case, the ability of exogenous ACTH to produce delivery, even when infused at physiological rates into the fetus, does not prove that this is the mechanism employed by the fetus.

It is clear from the circulating plasma ACTH concentrations in the fetus several weeks before delivery, that the fetal adrenal is exposed to the trophic action of ACTH from as early as 100 days gestation. The fetal adrenal can respond to exogenously administered ACTH in the intact animal at this stage, so it is probable that the in vivo plasma ACTH concentrations are inadequate to bring about the explosive growth of the adrenal at this time. Although the evidence available at the present time points to the conclusion that the development of adrenal cortical secretory capacity is the limiting factor in the mechanisms leading to delivery, the factor that controls this development remains a matter for speculation.

Fetal hypophysectomy will remove the basal trophic influence of ACTH on the fetal adrenal, causing adrenal hypoplasia and possibly leaving the adrenal insensitive to a signal other than ACTH. If the signal is a pituitary factor other than ACTH it will be absent in the hypophysectomised fetus. Prolactin is a possible steroidogenic factor which merits further attention [451]. Indeed, although the evidence is limited at the present time, fetal plasma prolactin concentrations show a greater tendency to rise over the critical period of adrenal development than do ACTH concentrations. Plasma concentrations of prolactin are usually low or unmeasurable between 100–120 days whereas measurable quantities are usually obtained in fetal plasma in late gestation. It is clear that we need a more detailed study of fetal plasma prolactin concentrations, relating them to fetal plasma cortisol concentrations. In addition, the effects of long-term prolactin infusions should be studied.

The signal to parturition may arise in another fetal tissue (Fig. 12.4). It is also possible that such an extra-pituitary tissue is under the control of the fetal pituitary, in which case fetal hypophysectomy would inevitably interfere with the signal.

There is a marked in vitro adrenal sensitivity to ACTH of adrenals removed from fetuses before 90 days. Subsequently there is a decrease in fetal adrenal sensitivity despite the high circulating plasma ACTH concentrations [592]. These observations suggest the possibility that some inhibitory factor of unknown origin is operating to suppress fetal adrenal sensitivity to ACTH during the period of 90–140 days. Removal of such an inhibitory effect might result in increased fetal adrenal cortisol secretion without any change in fetal plasma ACTH concentrations. The nature of such an inhibitory factor remains speculative. For several reasons it may not demonstrate itself in bioassay systems for ACTH concentrations. Firstly, it may show species specificity and lack any effect on the responsiveness of adrenal cells from species other than lambs. Secondly, it may have an effect locally, such as on adrenal blood flow, which would not manifest itself in investigations of in vitro sensitivity to ACTH. Thirdly, an inhibitory factor may exert its effects on a biochemical pathway that is not operative at the level of ascorbic acid depletion. It will be necessary to use a bioassay which involves all the steps in steroidogenesis and secretion such as the measurement of corticosteroid output.

Some of the various alternative hypotheses which can be put forward to explain the effects of fetal hypophysectomy or stalk section are shown in Fig. 12.4 [442]. The demonstration of a placental ACTH in the human by radioimmunoassay should be a reminder that stimuli to the adrenal need not come from the pituitary. Our data for fetal plasma ACTH concentrations post-hypophysectomy measured by bioassay are very low (<0.34 pg/ml) [Ratter, Rees, Thomas, Krane and Nathanielsz - unpublished observations]. In addition, radioimmunoassayable ACTH concentration is less than 10 pg/ml [330]. It should be recalled that the bioassay utilises adult guinea pig adrenal slices and not fetal

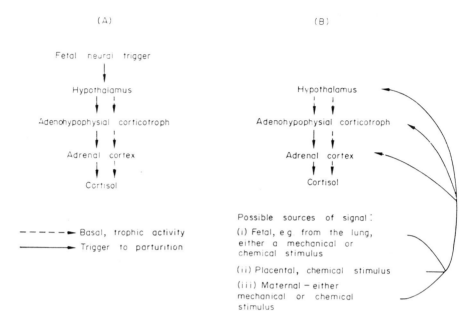

Fig. 12.4. Possible pathways of activity of the fetal hypothalamo–hypophysial–adrenocortical axis during gestation and parturition. (A) the simplest mechanism in keeping with the findings from fetal hypophysectomy and pituitary stalk section. (B) other possible mechanisms which would allow for loss of adrenal sensitivity after removal of fetal ACTH following fetal hypophysectomy.

sheep adrenal cells at the critical period of development. The radioimmunoassay data does not exclude a trophic factor of different molecular structure. Indeed to be absolutely certain of the nature and quantity of all adrenocorticotrophic factors in the plasma under any given experimental situation, a counsel of perfection would involve the use of a bioassay that utilises fetal lamb adrenal cells in the responsive period a day or so before delivery. In addition, the possibility of the existence of different molecular species will require some form of column chromatography of active plasma samples to determine the nature of the active agents.

One in vitro study of the growth and the development of corticosteroidogenesis has been carried out with human fetal adrenal cells [500]. Growth of the culture could be stimulated independently of steroidogenesis by the addition of fetal calf serum. Cyclic AMP stimulated steroidogenesis but not growth.

In summary, the nature of the stimulus to the increased production of cortisol by the fetal adrenal is still obscure. It is possible that the signal is programmed in the genetic code of the developing fetal adrenal. Alternatively, if the signal originates outside the adrenal, evidence is lacking as to its exact nature, source and timing. After the early successes in describing the rise in fetal lamb plasma cortisol and the demonstration of the ability of exogenous ACTH to initiate delivery, the experimental protocols which must be employed will need to be more rigorous in order to identify the location of the 'clock' that initiates the processes of parturition.

Measurement of fetal plasma ACTH and other pituitary hormone concentrations involve the problems of phasic changes in hormone concentration. In order to obtain a clear picture, it is necessary to undertake rapid multiple sampling. Endocrine gland ablations, especially hypophysectomy, remove several hormones and may have indirect effects via depression of function of other tissues such as the placenta. Two important experiments may throw some light on this system. Firstly, it would be of interest to selectively remove ACTH from the fetal plasma with a continuous infusion of antibody to ACTH. Secondly, now that normal ranges are available for fetal plasma ACTH, TSH, GH, prolactin and LH it should be possible to replace all or varying combinations of fetal pituitary hormones, at physiological concentrations, after fetal hypophysectomy or stalk section. Such an approach would be equivalent to the controlled ablation of individual series of pituitary cells, e.g. the thyrotrophs or the lactotrophs.

When we consider the question of possible signals from the conceptus at term, it should be borne in mind that there may be more than one signal and that some central integrating mechanism acts as a central co-ordinator to assess the insistence of the different inputs. Such a mechanism would have the biological merit of requiring several vital systems to signal their readiness for an extrauterine existence. Finally, a multifactorial system could more easily explain certain forms of premature labour. Pathological causes of premature delivery need not be an obstacle to an understanding of the normal, physiological mechanisms involved in parturition. Although pathology has played a great role in the development of the ideas we are about to discuss, it is clear that whilst some forms of pathological premature delivery yield useful information, other forms tell us nothing of the mechanisms involved in normal delivery [25]. A good example is abruptio placentae, in which placental separation may be followed by premature delivery. This tells us little about the physiology of the initiation of normal parturition.

Parturition is a multifactorial process involving several interconnected positive feedback loops (see, for example, Fig. 12.8). Both fetal and maternal factors are concerned. It is not always easy to distinguish between those processes which may play a role in the initiation of delivery and those whose role is to support labour once it has begun, and bring it to a successful conclusion.

Oxytocin is a good example of an agent whose effect is only supportive [222]. Maternal factors may play a role in the final timing of delivery in the sense that once labour has been initiated the actual time of delivery may be affected by various maternal factors. It is not surprising that there is a circadian rhythm in the incidence of human delivery with a peak incidence in the early hours of the morning [345]. Fetal factors may also be supportive. For example, experimental studies following brain aspiration in all the fetal rats in a litter show that, although parturition is not delayed, the time interval between the delivery of individual pups is increased [539a].

12.2.4 The output side – How does cortisol bring about delivery of the fetal lamb?

The various physiological roles of the glucocorticoids were discussed in Chapter 10. The role of cortisol as a potent inducer of enzyme activity has been considered in relation to the developing lung and the production of estrogens within the placenta. It has been suggested by some authors that cortisol may act in a 'permissive' fashion. It is the author's opinion that the term 'permissive' to describe the role of hormones has been used too loosely. The description permissive should be retained for those situations in which the hormone's presence is necessary to bring about some event, but other hormones or physiological systems are also operative. The permissive hormone is necessary for the event to take place but not sufficient by itself to bring about the change. Much has been said about the permissive actions of glucocorticoids. However, in their relation to the onset of parturition it is quite clear that fetal adrenal secretion of cortisol increases as term approaches [391, 442] and that the ability of exogenously administered glucocorticoid to bring about delivery is dose-dependent (Fig. 12.5). Dexamethasone has to be infused into the fetus at a rate of 0.06 mg/day, or greater, to induce delivery. In the human, dexamethasone is 26.7 times as potent as cortisol in terms of its glucocorticoid activity. Using this conversion factor (admittedly from the adult of a different species), our observed cortisol production rate of 9.17 mg/day (Table 10.3) is equivalent to 0.34 mg dexamethasone/day (Fig. 12.5). There are two main points to note from Fig. 12.5. Firstly, the ability of glucocorticoid to initiate delivery is dose-related. Secondly, there is an inevitable latency of 36–48 h before parturition occurs, even at high infusion rates.

Dexamethasone infusion to the ewe will produce delivery but the doses required are high, more than 4 mg/day [81, 235, 327]. It has been calculated that only up to 17% of the dexamethasone infused into the ewe crosses to the fetus [52]. However, Liggins [387a] observed that 4.0 mg dexamethasone/24 h was ineffective in the ewe whereas 0.06 mg/24 h (1.5%) of the maternal dose

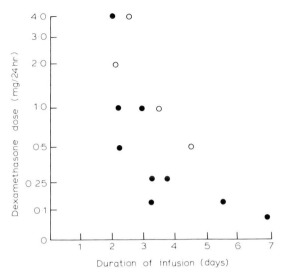

Fig. 12.5. Relationship between daily dose of dexamethasone infused intraperitoneally to the fetal lamb and latent period until parturition. Redrawn and reproduced with permission from Liggins [387].

was effective when infused into the fetus. It therefore appears that glucocorticoids administered from the fetal side of the placenta are more active than from the maternal side.

Two further experiments in which dexamethasone has been infused into the ewe need to be considered. Firstly, Bosc reports that dexamethasone injection to the hypophysectomised fetus does not produce delivery [82]. The ability of ACTH infusions to increase fetal adrenal glucocorticoid secretion in the hypophysectomised fetus has been demonstrated [330]. These ewes went into labour although labour was not normal. These three fetuses were hypophysectomised at 110 days gestation and ACTH infusion commenced at 125 days. A major problem with hypophysectomised animals is the possibility that the long-term effects of ACTH deficiency, both adrenal and extra-adrenal, may take time to return to normal when ACTH replacement is commenced. In addition various metabolic systems may return to normal at different rates. In the adult rat it has been demonstrated that there is a dissociation in the rate at which activity of various adrenal enzymes and adrenal corticosterone secretion are restored to normal by ACTH. In addition, hypophysectomy will alter the peripheral metabolism of glucocorticoids [132]. At the present time there are no published data reporting whether or not glucocorticoids infused into the hypophysectomised fetus will produce delivery. In one fetus, hypophysectomised cryosurgically at 126 days, infusion of 5.6 mg of cortisol/day was commenced 48 hours after surgery. Delivery occurred after 65.5 hours of cortisol infusion (Liggins and Nathanielsz, unpublished observations). Fetal plasma cortisol concentrations were 70 ng/ml at birth. Hypophysectomy was 84% complete as demonstrated by serial section of the fetal pituitary and fetal plasma growth hormone concentration had fallen to <6 ng/ml and fetal prolactin to <1.5 ng/ml within 72 hours of hypophysectomy. Liggins et al. [396] have demonstrated that hypophysectomy in which more than 70% of the pituitary is destroyed is followed by delay in the onset of parturition. If less than 70% of the pituitary is destroyed delivery occurs at term.

The second observation of great interest is that, when a single dose of 24 mg dexamethasone was given to fetal lambs at day 144, delivery occurred in a mean time interval of 30.1 hours, whereas when two doses of 24 mg were administered separately at day 144 and day 145 delivery occurred at 42.6 hours [83]. This was significantly later than in the group which received only one injection. The measurements of fetal and maternal plasma cortisol during these experiments should be noted. The high dose of dexamethasone used resulted in a suppression of fetal plasma cortisol concentrations at 6 hours after the dexamethasone injection but, by 24 hours the concentrations were back to control levels and by 32 hours the values in some animals were even higher than controls. The authors agree that this experiment is difficult to reconcile with the dose-related effects of dexamethasone infusion to the fetus (Fig. 12.5). However, the observation does show that feedback mechanisms may affect the results of administration of exogenous hormone.

The role of progesterone

Progesterone is capable of preventing uterine activity when administered to intact or ovariectomised pregnant females in several species [146, 147]. In general, species may be divided into two groups according to the source of progesterone in the latter stages of gestation [158, 498]. In some animals such as rat, rabbit and goat progesterone is produced almost entirely within the corpus luteum. In these animals, removal of the ovary at any stage of pregnancy leads to delivery. In the second group, which includes the sheep and the human, progesterone production in pregnancy is initially ovarian but eventually the placenta takes over the production of progesterone [234]. There may also be an adrenal contribution to circulating plasma progesterone concentrations [197]. The significance of the measurement of plasma concentrations of progesterone probably differs in these two groups, particularly after administration of exogenous progesterone. In the group in which progesterone production occurs in the corpus luteum, the progesterone has to reach the myometrial cell via the blood stream and therefore measurement of plasma progesterone concentrations is likely to be of use in following physiologically significant changes. In the group of animals in which progesterone is produced within the placenta, plasma concentrations may not closely reflect the local changes. Progesterone receptors exist on the myometrial cells [157]. Too little attention has been paid to the consideration of how the ultrastructure of the placenta, and the changes which occur in it at the time of delivery may influence passage of hormones between mother and fetus [323, 388, 537].

In the sheep, maternal plasma progesterone concentrations fall over the last 3 days or so of gestation [45, 57, 233, 234, 391, 554a]. This fall occurs after the increase in fetal plasma cortisol. It can be induced by the infusion of ACTH, dexamethasone or cortisol into the fetus. In general, maternal peripheral venous plasma concentrations of progesterone fall to about 1.0 ng/ml in both normal and glucocorticoid-induced delivery. However, parturition may occur spontaneously in the presence of maternal plasma progesterone concentrations of 3–5 ng/ml.

Administration of high doses of progesterone to the ewe (160 mg/day – the BPR for progesterone may be up to 50 mg/day [56]) does not always block the onset of uterine activity but may lead to an incoordinate type of uterine activity and difficulty with cervical dilation [285]. Liggins has demonstrated that 100 mg of progesterone administered to the pregnant ewe during the infusion of dexamethasone to the fetus, will not inhibit parturition although plasma and myometrial progesterone concentrations were maintained at normal concentrations [391]. It would be of interest to know the details of PG production and release in the ewes treated with 100 mg progesterone since 200 gm progesterone/day did not inhibit the rise of PG concentration in the endometrium and myometrium although it did prevent the release of PG into maternal plasma.*

In summary, progesterone is capable of exerting an inhibitory action on the myometrium in the sheep. Concentrations in twin pregnancies are higher than in ewes with a single fetus. These higher concentrations may act to offset the greater stimulatory effect of stretch of the uterus which occurs with multiple pregnancy. It has been mentioned previously that infusion of agents which are capable of inducing delivery in single fetuses may be ineffective at the same dose range in twin pregnancies. The local and systemic effects of progesterone produced by the non-infused twin may be of importance in such experiments.

---

* Progesterone may decrease the availability of estrogen receptor in the rat uterus [310]. In this study, the various layers of the uterus were not separated.

Progesterone does have an effect on single unit activity in the hypothalamus [40]. It is therefore probable that progesterone in maternal and fetal plasma affects hypothalamic neuroendocrine activity. However, the major mechanism of action of progesterone is probably via its influence on PG production and release. In addition, progesterone appears to be the substrate for estrogen production in the sheep [28, 534]. It is therefore possible that in normal delivery, the very processes which produce the rise in estrogen production themselves serve to remove progesterone. The failure of estrogens to rise in all instances that progesterone falls may be explained by only partial activity of the systems shown in Fig. 11.2. This problem highlights the multifactorial nature of parturition and shows how important it is to monitor all the various active principles when investigating individual experimental animals.

The role of estrogens

The administration of single injections of 20 mg stilbestrol to pregnant ewes at 136–141 days gestation leads to an increase in uterine activity after about 24 hours [283]. Eight of the 12 ewes in this study delivered but uterine activity regressed in the four ewes at the earliest gestational ages, and when these animals subsequently went into labour at term they developed cervical dystocia. There is no explanation currently available for the occurrence of cervical dystocia but we have observed a similar phenomenon in some ewes whose fetuses were infused with low doses of cortisol. It is apparent that normal labour requires an ordered series of interconnected events to occur in a correct sequence. Failure of any one of the normal changes to occur may lead to abnormal uterine activity. Several of the factors involved, for example, relaxin, have received very little study at the present time [520]. Estrogens will stimulate the production of PG in several different physiological situations [36, 75, 112, 113]. Estrogen administration to the pregnant ewe in late gestation will stimulate $PGF_{2\alpha}$ production and release in under 24 hours (Fig. 12.6). This time scale is compatible with the interpretation that estrogens produce their effects on myometrial activity via the production of PG. However, it should be noted that although maternal plasma estrogens may not rise when parturition is induced with glucocorticoids at an early gestational age (<124 days), PGF concentrations in maternal uterine vein plasma do rise as early as 110 days [391a].

The probable mechanism whereby fetal glucocorticoids induce the production of estrogens by the ovine placenta are discussed in Chapter 11. Since estrogen requires approximately 12–24 h to produce a marked increase in maternal venous PG concentration (Fig. 12.6), this leaves 12–24 h of the latent period in which glucocorticoids can produce delivery for glucocorticoids to

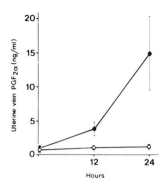

Fig. 12.6. Concentration of $PGF_{2\alpha}$ in uterine vein plasma after a single maternal injection of 20 mg of stilbestrol-in-oil with (○) and without (●) simultaneous intravenous infusion of 200 mg progesterone/24 h. Reproduced with permission from [388] Fig. 1.

increase estrogen production. Estrogens may have their action on PG production by affecting the stability of lysosomes [388]. This action may be mediated by an increased intracellular calcium concentration, as are other actions of estrogens [470].

The role of prostaglandins
The basic structure of the prostaglandins is shown in Fig. 12.7. This group of related molecules have potent oxytocic actions on the myometrial cell and are now considered by many workers to constitute the final common path of all agents which stimulate uterine contraction [148].

It has been demonstrated that increased lability or destruction of lysosomal membranes leads to a release of phospholipase A. This enzyme and, possibly other hydrolases releases PG precursors such as arachidonic acid from phos-

Fig. 12.7. The structure of PGE_1, PGE_2, $PGF_{1\alpha}$ and $PGF_{2\alpha}$.

pholipids within the cell. It has been demonstrated in the rabbit that maternal infusion of arachidonic acid precipitates delivery between 24–70 hours, whereas infusion of effective doses of $PGF_{2\alpha}$ will produce delivery in 36 ± 4 h [441]. Intraamniotic administration of arachidonic acid will also initiate labour in the human [399].

PG-induced contractions will lead to stimulation of oxytocin release via neuroendocrine reflexes and possibly, by a direct effect of PG on the neurohypophysis [243]. It has been observed that cervical vibration in the human expedites labour [85]. Cervical dilatation releases oxytocin in the human [206]. Manual dilatation of the cervix releases PG in the sheep [217], demonstrating the complex inputs to PG release in late pregnancy. These observations show that PG is involved in several processes which have a positive feedback component (Fig. 12.8). It has been mentioned before that the processes of parturition are multifactorial. It should now be noted that many of these factors are involved in positive feedback loops. It should also be noted that these positive feedback loops are interconnected. In short, parturition is a multifactorial system of interconnected positive feedback loops – at least on the output side. In addition, there are suggestions, discussed previously that the same multifactorial, interconnected positive feedback nature may operate on the input side. In general terms, positive feedback systems are rare in physiological processes. They have one inherent problem – they must be terminated by a specific event. They are, however, extremely useful in systems which have a definite termination, such as ovulation, or in parturition where, after the ex-

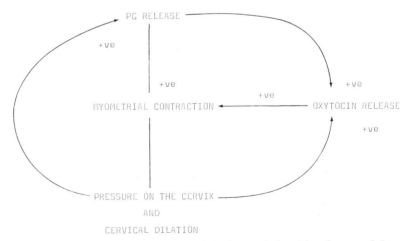

Fig. 12.8. Diagrammatic representation of the interrelationship of some of the events brought about by PG to demonstrate various positive feedback loops.

pulsion of the fetus, the various stimuli involved in the positive feedback are suddenly lost as a result of the activity of the system itself. In other words, the process of parturition is self-terminating and there are, therefore, no dangers in the positive feedback, providing no obstruction occurs. Indeed, once commenced, it is of great importance that delivery should occur as rapidly as is compatible with good phases of uterine relaxation to permit adequate placental blood flow.

The role of the cyclic nucleotides in myometrial contraction
Although it was suggested earlier that, according to some workers, PG is the final common path for all agents which act on the myometrial cell, it is clear that this description depends upon the level at which the observer is investigating the activity of this contractile cell. It appears that PG and oxytocin stimulate myometrial contraction by a mechanism which decreases tissue levels of cyclic AMP [66]. This hypothesis has been further modified to include the idea that an inverse relationship exists between cyclic AMP and cyclic GMP. According to this theory, cell function is controlled by a change in the ratio of these two nucleotides, not simply by changes in cyclic AMP concentration alone [372, 373, 246]; a high cyclic GMP/cyclic AMP ratio is associated with contraction of smooth muscle, whereas a low ratio is associated with relaxation.

Experimental evidence regarding these hypotheses has been produced generally in the rat. Oxytocin and $PGF_{2\alpha}$ increase myometrial concentrations of cyclic GMP without affecting cyclic AMP concentrations [372, 373]. In addition, in other smooth muscle in which $PGF_{2\alpha}$ and PGE_2 have opposite effects, these two prostaglandins also have opposite effects on the ratio of cyclic nucleotides [177]. However, the intra-arterial infusion of both cyclic nucleotides or theophylline, which will inhibit the destruction of endogenous cyclic nucleotides, will completely inhibit $PGF_{2\alpha}$-induced contractions and considerably reduce both oxytocin-induced and spontaneous uterine contractions in the rat and rabbit [97, 320, 321]. The difference in the observed effects in both the rat and rabbit between $PGF_{2\alpha}$- and oxytocin-induced contractions suggests that, if oxytocin causes myometrial contraction via the release of $PGF_{2\alpha}$, this mechanism may differ from the action of $PGF_{2\alpha}$ acting alone. The suggestion has been made that the role of $PGF_{2\alpha}$ in oxytocin-induced contractions is facilitatory rather than obligatory [569]. It is, however, clear that the interpretation of results from exogenous administration of metabolites which are normally intracellular, must be approached with caution. However, this approach can yield useful data, especially when it is recalled that measurements of tissue concentrations of such fundamental cellular materials as prostaglandins and cyclic nucleotides have often to be made in a very heterogenous cell population. In addition, even within a homogenous cell population, the prostaglandins and cyclic nucleotides subserve a large number of the many functions of the cell. This fact complicates interpretation of changes in their concentrations in active tissues.

12.3 The progress of pregnancy following fetectomy in experimental animals

The conclusion that the fetus initiates the processes discussed in the early part of this chapter is often criticised as being contrary to the evidence produced from experiments in which the fetus is destroyed in utero or removed whilst leaving the placenta intact. When referring to several of the reports in which fetectomy was performed, it is often claimed – sometimes contrary to the claim of the authors themselves – that delivery was normal and took place at term. Unfortunately these experiments were often conducted before it was possible to measure maternal hormone concentrations and it is impossible to know whether parturition was endocrinologically within the normal limits. In addition, such experiments drastically change the volume and nature of the uterine contents. Thirdly, it is clear that there are differences between polytocous and monotocous species. For example, in polytocous species litter size may affect the duration of pregnancy [170]. Finally, a careful study of the actual experimental regimes employed in these various experiments and the actual claims made by the authors should be made.

There are three studies to which we may refer in the rat and mouse [361, 448, 512]. In both of the studies in the rat, fetectomy was performed over a very wide range of gestational ages 14–21 days [361] and 9–13 days [512]. In the former study, exploratory laparotomy was performed on day 22 to ascertain whether the placentae were still in situ. Delivery then occurred at day 23 or 24 (normal delivery in this colony was 23 ± 1 days). Ten experiments were conducted over this entire range but the spread of gestational ages at which each individual experiment was performed was not noted. It is always possible that delivery occurring 1–2 days after laparotomy on day 23 was influenced by the stress of surgery; no control was performed for this possibility.

In the second study [512], twelve pregnant rats underwent total fetectomy between day 9 and 13 of pregnancy. Delivery of placentae occurred 8–12 days later. The range of gestational ages at delivery was not given. The authors conclude, 'This means that the length of pregnancy was not considerably altered by the removal of the embryo. We think that these experiments justify the conclusion that the placenta determines the length of pregnancy, either by its effect on the corpora lutea, or by its own progestin production or – which is more likely – by both these means.' The conclusions are justifiably cautious and it is likely that the duration of pregnancy was not 'considerably altered'. In a gestation period of 21 days, one day represents 5% and it is not possible to use each animal as its own control. It would be necessary to use much larger numbers and a more homogenous group to produce a statistically significant effect. In the mouse [448], twenty animals, fetectomised in groups of 5 on days 12, 13, 14 and 15 of pregnancy, delivered on day 16–23. Normal parturition occurred mostly on days 19, 20 and 21 in this colony, although some unoperated animals did deliver on days 22 and 23.

The classical experiment of Van Wegenen and Newton [565] is the one most commonly quoted in relation to the effect of fetectomy in the primate. These authors fetectomised eleven rhesus monkeys at 70–157 days gestation and delivery occurred at 114–184 days. The normal gestation length in this species is 168 days. A more recent and detailed study is that of Lanman et al. [378] in which measurements of maternal plasma progesterone concentration were also made. In contrast to the findings of Van Wagenen and Newton, in 5 out of 8 pregnant rhesus monkeys fetectomised at 80–140 days, delivery occurred past the upper limit of normal gestation length. One placenta was delivered before the normal range of times and two were delivered at times which could be considered normal. Maternal plasma progesterone concentrations were maintained following fetectomy. However, when ovariectomy was performed following fetectomy in three animals, delivery occurred within 3–8 days. A third study reports the progress of pregnancy in four pregnant rhesus monkeys following umbilical cord ligation at 101–114 days gestation. Abortion occurred after 3 days in one monkey and after 16–21 days in the other three [84].

The duration of pregnancy and the progress of labour is very variable even within one species [377]. It is unlikely that experiments involving such a radical procedure as fetectomy will prove very valuable in our attempts to pinpoint the source of the normal signal to the initiation of parturition. They may, however, throw light on the interrelationship of fetus, placenta and ovary. The normal progress of labour can be affected by the aspiration of the fetal brain in the rat [540]. It should therefore be borne in mind that fetectomy will have effects on both the initiation and successful completion of parturition.

In concluding this section I shall quote from the summary of the paper by Van Wagenen and Newton [565]. 'The placenta remained in the uterus for the normal duration of pregnancy (defined as a time compatible with the expectation of a viable baby). Onset of labour and delivery of the placenta were normal. The placenta was spontaneously expelled after remaining in the uterus as long as three months in the absence of the fetus'. This conclusion is very different from the usual claims made when referring to this paper that the presence of the fetus has no influence on the timing of delivery or that delivery of the placenta occurs at term following fetectomy. In any event these conclusions cannot be drawn from the more recent studies [84, 378], the second of which speaks in favour of the fetus initiating parturition since the majority 5 out of 8 placentae were delivered post-term. The study by Lanman et al. [378] provides both histological and endocrinological data and shows clearly that, without such observations, experiments involving fetectomy are entirely inconclusive and provide little more than curious results.

12.4 The initiation, maintenance and successful completion of parturition in the human and sub-human primate

12.4.1 Historical background

The possibility that the human fetus has an influence on the timing of the onset of parturition was suggested by the reports of post-maturity in pregnancies in which the fetus had varying forms of anencephaly or other cranial abnormalities [26, 62, 133, 406, 421, 484, 558]. Hypophysectomy of the fetal rhesus monkey leads to delay in the onset of parturition [122]. This observation suggests a similarity to the observations described in the sheep and is of considerable importance in view of the fact that the small amount of endocrinology data available from the human anencephalic shows quite clearly that they are a very heterogenous group [20, 590]. For example, in two anencephalic neonates, cord blood plasma concentrations of ACTH were 51 and 71 pg/ml [590]. The normal range was 20–1330 pg/ml. The variation in the degree of pituitary function may account for the failure of some authors to demonstrate a prolongation of gestation in cases of anencephaly [541]. The human anencephalic probably has more in common with the stalk-sectioned experimental animal than the hypophysectomised animal. The presence of varying amounts of pituitary tissue [508], dislocated from its normal connection with the hypothalamus shows some similarities to the situation in the brain of the abnormal sheep fetuses who had been exposed to the toxoid of *Veratrum californicum* at the critical phase of development. In cases of adrenal hypoplasia in the human without any outward appearance of hypopituitarism, parturition is delayed beyond term [457]. Results from bilateral fetal adrenalectomy at 100–120 days gestation in the rhesus monkey are inconclusive. Although 6 of the 8 animals delivered between 143 and 167 days (normal term = 168 days), the remaining two delivered 22 and 49 days post-mature [428a].

12.4.2 *The input to the fetal adrenal in the primate*

The many problems associated with the investigation of the control of the fetal sheep adrenal have been discussed at length. It is very unlikely that the situation is any simpler in the primate.

Human cord blood plasma ACTH concentrations have been measured by radioimmunoassay. At 12–19 weeks they are 249 pg/ml; 20–34 weeks, 234 pg/ml and at 35–42 weeks, 143 pg/ml. There was a significant fall in ACTH from 20–34 weeks to 35–42 weeks [590]. This data is in keeping with the failure to demonstrate a rise in fetal ACTH in the sheep to precede the rise in fetal plasma cortisol. Surprisingly, there was no difference in cord blood ACTH levels related to the mode of delivery. Values were 145 pg/ml for spontaneous vaginal delivery, 126 pg/ml for caesarian section in labour, 136 pg/ml for caesarian section prior to labour and 121 pg/ml for vaginal delivery following oxytocin induction. These results contrast sharply with the sheep in which vaginal delivery caused an increase in ACTH in cord blood from about 500 pg/ml up to as high as 3000 pg/ml. In the human fetal plasma ACTH concentrations do not appear to be higher than maternal plasma ACTH concentrations [19, 64, 590], as is the case in the sheep. There is no relationship between maternal and fetal plasma ACTH concentrations. It has been demonstrated that the fetal monkey adrenal is responsive to ACTH in the second half of gestation [363, 365]. There is also evidence for a placental ACTH [241, 485].

In the human anencephalic the fetal zone of the adrenal regresses prematurely. A similar regression of the fetal zone generally occurs at the time of delivery. Several explanations have been proposed for the normal regression of the fetal cortex. One possibility is obviously the withdrawal of a placental factor. hCG has been shown to produce ultrastructural changes in the fetal adrenal in vitro [328]. In vivo studies with anencephalic fetuses have demonstrated a minimal effect of ACTH and no stimulation by hCG [301]. However, the acute responsiveness of the adrenal of the anencephalic may well be diminished. Finally, in vitro studies show that the fetal definitive cortex can produce glucocorticoids [514] and the relative roles of the fetal zone and the definitive zone must remain uncertain at the present time.

12.4.3 *The output side from the fetal adrenal*

Fetal secretion of cortisol in the primate

The various forms of evidence which suggest that fetal adrenal secretion of cortisol increases gradually before term both in the monkey and the human are discussed in Chapter 10. The most convincing evidence comes from measurements of serial amniotic fluid concentrations [434]. These show that the increase is slower than in the sheep, occurring over several days, but that the magnitude of the rise is similar (see Figs. 9.3 and 10.9).

The human fetal adrenal differs from the sheep fetal adrenal in that the human fetal adrenal lacks the 3β-ol-hydroxysteroid dehydrogenase and therefore has to use progesterone as a substrate for cortisol production [532]. A further difference is that the primate adrenal does possess 17-hydroxylase and 17,20-lyase activity and is therefore capable of producing C19 steroids (Fig. 11.6) whereas the sheep adrenal is not. In contrast to the sheep placenta, the primate placenta does not show 17-hydroxylase or 17,20-lyase activity. However, in the normal fetus, although cortisol is produced during the later stages of gestation, steroids, derived from pregnenolone (the Δ^5 pathway) are present in larger quantities than those produced by the Δ^4 pathway from progesterone [492].

In summary, at the present time the differences between the sheep and the primate in terms of fetal cortisol production appear to be related to the nature of the substrate used and the time course of the increase in fetal cortisol secretion. However, the primate studies have been carried out on amniotic fluid and no such studies have been carried out in the sheep.

Glucocorticoid administration in the ovine and primate pregnancy Dexamethasone administration either directly to the fetus or to the pregnant ewe will precipitate delivery. Parturition is generally accompanied by the normal changes in maternal plasma estrogen and progesterone concentrations. The experimental administration of dexamethasone in the pregnant primate has not been carried out by so many different routes or at the wide range of doses and gestational ages investigated in the sheep. This is of importance in connection with the ideas discussed in Chapter 9 that glucocorticoid may be exerting various feedback and induction effects at several levels, including the fetal adrenal.

In a very detailed study, Challis and co-workers were unable to precipitate delivery by the administration of 8 mg dexamethasone i.m. daily to pregnant rhesus monkeys from about 150 days gestation until delivery which occurred at 172 days ± 4.3 (S.D.). Normal gestation length was 174.6 ± 5.8 for the controls [110, 111]. A similar dose of dexamethasone on a weight for weight basis would be effective in the sheep [81, 235].

Dexamethasone crossed the placenta (as it has been shown to do in several species) and produced adrenal atrophy in the fetus. There was also a drop in maternal plasma estrogen concentrations to 30% of the level before treatment. This was accompanied by a fall in maternal testosterone and androstenedione to about 60% of control concentrations. The normal increase in maternal estrogens which generally occurs in the last 10 days of gestation did not occur in the dexamethasone-treated animals. The authors therefore conclude that the marked rise in plasma estrogens which occurs in the last 10 days of pregnancy is a function of fetal production of C19 precursors.

This experiment raises two separate but related questions. Firstly, why did the dexamethasone not induce premature delivery as it would have done in the sheep? Secondly, why did delivery occur at the expected time in the absence of an increase in maternal estrogen concentration? The first question is easier to answer than the second and depends upon the different source of estrogen precursors in the sheep and the human.

In the sheep, high doses of dexamethasone will almost certainly inhibit fetal cortisol production (see data in Fig. 9.8 of suppression of fetal ACTH by exogenous cortisol). However, the estrogen precursors are produced in the placenta. Their production is not inhibited by dexamethasone and they do not become rate-limiting. However, in the primate, dexamethasone suppresses both the maternal and the fetal adrenal, thus inhibiting production of C19 precursor for the placental synthesis of estrogens which therefore become rate-limiting. The various components of the feto–placental unit are therefore playing very different roles in the sheep and the primate. Nevertheless there is interaction between the fetus (in particular the adrenal) and the placenta in both instances.

The administration of synthetic glucocorticoids in an attempt to induce premature delivery in the human has led to conflicting results. Despite failures by several groups, Mati et al. [409] produced significantly earlier delivery in women with post-maturity when betamethasone was injected into the amniotic cavity when compared with controls. This study may have succeeded where others failed because they used women who were post-mature as subjects. Negative findings in this field are difficult to interpret in view of the several systems which must be responsive, the problems of transplacental passage of dexamethasone and time-course of action of single injections. Unfortunately, no data are available for the changes in maternal steroid hormones in these induced deliveries.

The role of progesterone
The suggestion that progesterone plays a role in blocking uterine contractions, possibly by affecting PG release, has been discussed in relation to the experimental work in the sheep. In the human, maternal plasma progesterone concentrations do not appear to change prior to delivery, although as discussed in Chapter 11, there may be changes in the estrogen : progesterone ratio. At the present time it can be said that progesterone would be capable of exerting an

inhibitory action on the myometrial cell and thereby exert an important role in the maintenance of pregnancy. Whether progesterone withdrawal is important in the primate as a prerequisite to labour is unlikely. Parturition, which eventually follows both fetal death and dexamethasone administration in the rhesus monkey, is not preceded by a fall in maternal plasma progesterone concentration. In the light of these observations it may be concluded that, as in the sheep, progesterone withdrawal is not a necessary prerequisite for the onset of labour.

The role of estrogens
Similar problems exist in the primate when we attempt to assess the role of estrogens. Estrogens undoubtedly stimulate uterine growth and blood flow in the primate. They may also stimulate PG production. However, dexamethasone administration to the pregnant rhesus monkey does not produce premature delivery; delivery occurs after about the normal gestation period but without the normal rise in maternal plasma estrogens over the last 10 days of gestation. It should be recalled that delivery can occur in the sheep without any rise in maternal plasma estrogen concentrations.

The role of prostaglandins
The concentration of PG in fetal and maternal tissues was discussed in Chapter 11. It is clear that PG does play a role in uterine contraction in the primate [108]. In addition, amniotic fluid $PGF_{2\alpha}$ and $PGF_{2\alpha}$ metabolite concentrations rise before delivery in the rhesus monkey [114a]. Exogenous PG will produce uterine contraction more readily in the human than the sheep and this may be related to the differences in the metabolism of PG by the uterus and fetal membranes in these two species. Chronic administration of indomethacin delays parturition [453a]. PG also probably plays a role in stimulating cervical dilation (Table 12.1).

12.5 Future work

Throughout this book the interplay of different endocrine and metabolic factors at different stages of development has been discussed. In the chronically catheterised sheep fetus with indwelling vascular catheters it is now possible to monitor a vast array of physiological and endocrine functions. The best examples of such interactions which have been cited are drawn from the adrenal and thyroid axes. It is quite clear that the physiological significance of the changes in the thyroid axis will only be clarified when we at least know the detailed changes which occur in TSH, T_4, T_3 and rT_3 concentrations, together with an

Table 12.1.
Some of the important factors that may play a role in the maintenance of pregnancy and the initiation of parturition in the sheep on the output side as defined in 12.2.4.

Maternal factors	Fetal factors	Action
Progesterone	Progesterone – probably not an important factor because of inactivation by 20α-HSD	1. Decreases myometrial activity 2. Acts as a substrate for estrogen production 3. Inhibits the release of PG
Estrogens	Estrogens	1. Promote uterine and blood flow 2. Stimulate PG synthesis 3. Decrease oxytocin threshold of myometrium – probably via PG
Prostaglandin (PG)		1. Has an important oxytocic effect – may even be the final common path for all oxytocic agents 2. Stimulates oxytocin release 3. Produces cervical dilatation
	Prostaglandins	1. Produces closure of the ductus arteriosus and has effects on other parts of the fetal vasculature 2. PG presents on the fetal side of the placenta can exert an oxytocic effect
Oxytocin	Oxytocin	Oxytocin present on both sides of the placenta can exert an oxytocic effect
Cyclic nucleotides	These are discussed in relation to the coupling of oxytocin and PG to uterine contraction	
AVP		1. AVP may be an ACTH-releasing factor 2. At high concentrations AVP can exert an oxytocic effect. It is uncertain whether this can occur at physiological concentrations

assessment of how varying the concentrations of individual hormones affects the secretion and metabolism of the other hormones in the axis. There are other examples of these interactions, some of which have been quoted in this book.

Future experimental work on the fetal sheep preparation will be much more exacting than what has gone before. Good animal husbandry and good vascular catheterisation will remain important basic prerequisites to success but no more than a starting point. Much has been made of the difficulties of fetal sheep catheterisation by those who wish to surround it by mystique. The actual surgical procedures can be carried through by any recently qualified surgeon with three or four weeks introduction to the discipline of fetal work. The success of this fascinating field depends on an interdisciplinary approach. It requires the application of modern methods of endocrine measurement, coupled to physiological techniques such as EEG, ECG and especially, uterine contraction records. In addition, the observations of diurnal patterns in the fetus mean that the maintenance of patency of the catheters and the 24-hour monitoring of the fetus and mother demand that each preparation receives the full 24-hours a day care that a patient in a hospital intensive care unit would receive.

Appendix

Methods available for the calculation of the production rate of hormones by the fetus

There are various methods available for the calculation of hormone secretion rates by the endocrine glands of the fetus. Blood production rates (BPR) in the fetus must take into account also the transfer of hormones across the placenta between the maternal and fetal compartments.

1. Removal of the endocrine gland under observation

Surgical removal, or total ablation, of the endocrine gland under observation can be followed by replacement of exogenously administered hormone. If the hormone concentrations during replacement can be made equal to those before removal of the gland, the BPR in the fetal compartment equals the rate at which the hormone is being administered. This approach is not very accurate when used to investigate the secretion of hormones whose levels normally fluctuate rapidly.

2. Isotopic techniques

The use of very small amounts of radioactively labelled hormone which mix completely with endogenously secreted hormone molecules is now the method of preference. These techniques minimise the disturbance to fetal endocrinology caused by the experimental investigation of the prevailing secretion rate. They also avoid changes which may occur in the peripheral metabolism of the hormones under investigation following removal of the gland.

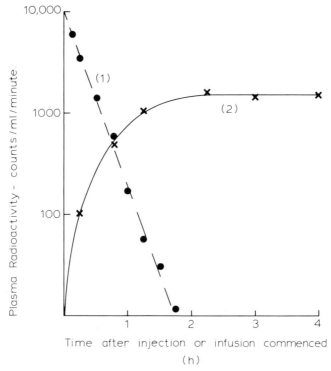

Fig. A.1. Blood radioactivity after the injection of a single bolus of labelled steroid intravascularly (1) or the constant infusion of labelled steroid (2) in the fetal sheep. These data are diagrammatic but show how kinetic calculations are conducted. In this instance the steroid under investigation is considered to be distributed in a single compartment and the disappearance curve (1) follows a single exponential course.

There are two main techniques:

(a) Single injection of the isotopically labelled hormone followed by observation of the disappearance of the labelled hormone from the blood. If the radioactivity in the blood is plotted on semilogarithmic graph paper the half-life ($t_{\frac{1}{2}}$) of the hormone in blood can be measured (Fig. A1). The fractional turnover rate, k, of the hormone pool can then be calculated:

$$k = \frac{\log_e 2}{t_{\frac{1}{2}}}$$ (% per minute, hour or day where the half life, $t_{\frac{1}{2}}$ is given in minutes, hours or days, respectively)

If the amount of radioactivity injected is R (counts/min), the volume of distribution of the hormone, V (litres), can be calculated by extrapolation back to zero time of the disappearance curve after full equilibration of the label throughout the volume of distribution. If the theoretical initial radioactivity is I (counts/min), then $V = R/I$. If [C] is the concentration of endogenous, non-labelled hormone in plasma in µg/litre, then the total body content of hormone is $V \times [C]$.

The MCR $= V \times k$

The BPR for the hormone $= V \times [C] \times k$ µg/day $=$ MCR $\times [C]$ (this amount is often expressed/Kg body weight).

(b) *Continuous infusion techniques.* The radioactive form of the hormone may be infused into the fetus at a constant rate. The ratio of the rate of infusion of radioactive hormone to the final concentration of radioactive hormone will be equal to the ratio of the rate at which non-labelled hormone enters the blood (BPR) to the concentration of non-labelled hormone in the blood [C].

The metabolic clearance rate (MCR) of a hormone from blood is that volume of blood which would have to be completely and irreversibly cleared of the hormone per unit time to account for the amount of hormone removed from the whole vascular compartment. It is obviously a theoretical concept which, however, like the concept of renal clearance, describes the amount of the molecule under investigation which is removed from the circulation per unit time. If the plasma hormone concentration remains stable, this amount of hormone must be replaced per unit time.

Thus, BPR $=$ MCR (1/24 h) $\times$ plasma concentration of hormone (µg/l), and is expressed in µg/day

$$\text{MCR (ml/min)} = \frac{\text{Rate of infusion of radioactive hormone (counts/day)}}{\text{Final concentration of radioactive hormone at equilibrium (counts/l)}}$$

Obviously the infusion solution and the plasma sample for measuring the radioactivity at equilibrium must be counted in the radioactive counter for the same length of time. From the point of view of calculating MCR, the endocrine gland and the pump infusing the radioactive hormone are fulfilling similar roles. The gland is secreting unlabelled hormone which is diluted in the animal's hormone pool; the infusion pump is infusing labelled hormone which is diluted in the same pool. It is fundamental to this methodology that labelled and unlabelled hormone are metabolised in the same way within the hormone pool.

Effect of placental transfer

If placental transfer of hormone occurs, labelled hormone infused into the maternal circulation will enter the fetal hormone pool. This problem can be investigated in two ways. Two different labelled forms of the hormone may be infused at the same time ([^{14}C]cortisol to the fetus and [^{3}H]cortisol to the ewe, for example) or two experiments can be carried out, allowing time for the labelled hormone to be cleared from the animal. This is possible in experiments with steroids which are very rapidly cleared.

The extent of transfer of cortisol between the two cortisol pools can be calculated. The BPR for cortisol within each compartment represents the adrenal secretion rate and the transfer added together. It is therefore possible to calculate the secretion rate by subtracting that quantity of hormone which has been transferred across the placenta from the blood production rate [257].

Bibliography

1. Aarskog, D. (1965) Cortisol in the newborn infant. Studies on cortisol concentration and protein-binding in umbilical cord plasma and maternal plasma at delivery, and on cortisol production and metabolism in newborn, full-term and premature infants, with special reference to newborn infants of diabetic mothers. *Acta Paediat. Scand.* suppl. 158, 1-91.
2. Abdul-Karim, R.W., Nesbitt, R.E.L. Jr., Drucker, M.H. and Rizk, P.T. (1971) The regulatory effect of estrogens on fetal growth. 1. Placental and fetal body weights. *Am. J. Obstet. Gynecol.* 109, 656-661.
3. Abramovich, D.R. and Rowe, P. (1973) Foetal plasma testosterone levels at mid-pregnancy and at term: relationship to foetal sex. *J. Endocrinol.* 56, 621-622.
4. Abrams, R.L., Grumbach, M.M. and Kaplan, S.L. (1971) The effect of administration of human growth hormone on the plasma growth hormone, cortisol, glucose and free fatty acid response to insulin: evidence for growth hormone autoregulation in man. *J. Clin. Invest.* 50, 940-950.
5. Adams, J.H., Daniel, P.M. and Prichard, M.M. (1963) The effect of stalk section on the volume of the pituitary gland of the sheep. *Acta Endocrinol.* 43, suppl. 181, 1-27.
6. Albano, J.D.M., Jack, P.M., Joseph, T., Gould, R.P., Nathanielsz, P.W. and Brown, B.L. (1976) The development of ACTH-sensitive adenylate cyclase activity in the foetal rabbit adrenal: a correlated biochemical and morphological study *J. Endocrinol.* 71, Part III, December.
7. Alexander, D.P., Bashore, R.A., Britton, H.G. and Forsling, M.L. (1974) Maternal and fetal arginine vasopressin in the chronically catheterised sheep. *Biol. Neonate* 25, 242-248.
8. Alexander, D.P., Britton, H.G., Buttle, H.L. and Nixon, D.A. (1972) Prolactin in the sheep foetus. *Res. Vet. Sci.* 13, 188-189.
9. Alexander, D.P., Britton, H.G., Cameron, E., Foster, C.L. and Nixon, D.A. (1973) Adenohypophysis of foetal sheep: correlation of ultrastructure with functional activity. *J. Physiol.* 230, 10-12P.
10. Alexander, D.P., Britton, H.G., Forsling, M.L., Nixon, D.A. and Ratcliffe, J.G. (1971) The release of corticotrophin and vasopressin in the foetal sheep in response to haemorrhage. *J. Physiol.* 213, 31-32P.

11. Alexander, D.P., Britton, H.G., Forsling, M.L., Nixon, D.A. and Ratcliffe, J.G. (1971) The concentrations of adrenocorticotrophin, vasopressin and oxytocin in the foetal and maternal plasma of the sheep in the latter half of gestation. *J. Endocrinol.* 49, 179-180.
12. Alexander, D.P., Britton, H.G., Forsling, M.L., Nixon, D.A. and Ratcliffe, J.G. (1973) Adrenocorticotrophin and vasopressin in foetal sheep and the response to stress. In: *The Endocrinology of Pregnancy and Parturition.* Ed.: Pierrepoint, C.G. (Tenovus Library Series, Alpha Omega Alpha, Cardiff) pp. 112-123.
13. Alexander, D.P., Britton, H.G., Nixon, D.A., Cameron, E., Foster, C.L., Buckle, R.M. and Smith, F.G. (1973) Calcium, parathyroid hormone and calcitonin in the foetus. In: *Foetal and Neonatal Physiology.* Proc. Sir Joseph Barcroft Centenary Symp. Eds.: Comline, R.S., Cross, K.W., Dawes, G.D. and Nathanielsz, P.W. (Cambridge University Press) pp. 421-429.
14. Alexander, D.P., Britton, H.G., Forsling, M.L., Nixon, D.A. and Ratcliffe, J.G. (1974) Pituitary and plasma concentrations of adrenocorticotrophin, growth hormone, vasopressin, and oxytocin in fetal and maternal sheep during the latter half of gestation and the response to haemorrhage. *Biol. Neonate* 24, 206-219.
15. Alexander, D.P., Britton, H.G., Nixon, D.A., Ratcliffe, J.G. and Redstone, D. (1973) Corticotrophin and cortisol concentrations in the plasma of the chronically catheterised sheep fetus. *Biol. Neonate* 23, 184-192.
16. Alexander, D.P., Britton, H.G., James, V.H.T., Nixon, D.A., Parker, R.A., Wintour, E.M. and Wright, R.D. (1968) Steroid secretion by the adrenal gland of foetal and neonatal sheep. *J. Endocrinol.* 40, 1-13.
17. Alexander, D.P., Forsling, M.L., Martin, M.J., Nixon, D.A., Ratcliffe, J.G., Redstone, D. and Tunbridge, D. (1972) The effect of maternal hypoxia on fetal pituitary hormone release in the sheep. *Biol. Neonate* 21, 219-228.
18. Allen, C. and Kendall, J.W. (1967) Maturation of the circadian rhythm of plasma corticosterone in the rat. *Endocrinology* 80, 926-930.
19. Allen, J.P., Cook, D.M., Kendall, J.W. and McGilvra, R. (1973) Maternal-fetal ACTH relationship in man. *J. Clin. Endocrinol. Metab.* 37, 230-234.
20. Allen, J.P., Greer, M.H., McGilvra, R., Castro, H. and Fisher, D.H. (1974) Endocrine function in an anencephalic infant. *J. Clin. Endocrinol. Metab.* 38, 94-98.
21. Anderson, A.B.M., Flint, A.P.F. and Turnbull, A.C. (1974) Mechanism of action of glucocorticoids in induction of ovine parturition: effect on placental steroid metabolism. *J. Endocrinol.* 61, xxxvi-xxxvii.
22. Anderson, A.B.M., Flint, A.P.F. and Turnbull, A.C. (1975) Mechanism of action of glucocorticoids in induction of ovine parturition: effect on placental steroid metabolism. *J. Endocrinol.* 66, 61-70.
23. Anderson, A.B.M., Pierrepoint, C.G., Griffiths, K. and Turnbull, A.C. (1972) Steroid metabolism in the adrenals of fetal sheep in relation to natural and corticotrophin-induced parturition. *J. Reprod. Fertil.*, suppl. 16, 25-37.
24. Anderson, A.B.M., Pierrepoint, C.G., Turnbull, A.C. and Griffiths, R. (1973) Steroid investigations in the developing sheep foetus. In: *The Endocrinology of Pregnancy and Parturition.* Ed.: Pierrepoint, C.G. (Tenovus Library Series, Alpha Omega Alpha, Cardiff) pp. 23-39.
25. Anderson, A.B.M., Laurence, K.M., Davies, K., Campbell, H. and Turnbull, A.C. (1971) Fetal adrenal weight and the causes of premature delivery. *J. Obstet. Gynaecol. Br. Commonw.* 75, 481-488.

26. Anderson, A.B.M. and Turnbull, A.C. (1973) Comparative aspects of factors involved in the onset of labour in ovine and human pregnancy. *Mem. Soc. Endocrinol.* 20, 141-162.
27. Arai, Y. and Gorski, R.A. (1968) Critical exposure time for androgenization of the developing hypothalamus in the female rat. *Endocrinology* 82, 1010-1014.
28. Ash, R.W., Challis, J.R.G., Harrison, F.A., Heap, R.B., Illingworth, D.V., Perry, J.S. and Poyser, N.L. (1973) Hormonal control of pregnancy and parturition; a comparative analysis. In: *Foetal and Neonatal Physiology*: Proc. Sir Joseph Barcroft Centenary Symp. Eds. Comline, R.S., Cross, K.W., Dawes, G.S. and Nathanielsz, P.W. (Cambridge University Press) pp. 551-561.
29. Attal, J. (1969) Levels of testosterone, androstenedione, estrone and estradiol 17β in the testes of fetal sheep. *Endocrinology* 85, 280-289.
30. Austin, C.R. and Short, R.V. (1972) *Embryonic and Fetal Development.* (Cambridge University Press).
31. Avery, M.E. (1975) Pharmacological approaches to acceleration of fetal lung maturation. In: *Perinatal Research, Br. Med. Bull.* 31 (1) 13-17.
32. Azizi, F., Vagenakis, A.G., Bollinger, J., Reichlin, S., Braverman, L.E. and Ingbar, S.H. (1974) Persistent abnormalities in pituitary function following neonatal thyrotoxicosis in the rat. *Endocrinology* 94, 1681-1688.
33. Baker, F.D. and Kragt, C.L. (1969) Maturation of the hypothalamic–pituitary–gonadal negative feedback system. *Endocrinology* 85, 522-527.
34. Bakke, J.L., Gellert, R.J. and Lawrence, N. (1970) The persistent effects of perinatal hypothyroidism, on pituitary, thyroid and gonadal functions. *J. Lab. Clin. Med.* 76, 25-33.
35. Bakke, J.L. and Lawrence, N. (1966) Persistent thyrotrophin insufficiency following neonatal thyroxine administration. *J. Lab. Clin. Med.* 67, 477-494.
36. Barcikowski, B., Carlson, J.C., Wilson, L. and McCracken J.A. (1974) The effect of endogenous and exogenous estradiol-17β on the release of prostaglandin $F_{2\alpha}$ from the ovine uterus. *Endocrinology* 95, 1340-1349.
37. Barnes, C.M., Warner, D.E., Marks, S. and Bustad, L.K. (1957) Thyroid function in foetal sheep. *Endocrinology* 60, 325-328.
38. Barnes, R.J. (1976) Water and mineral exchange between maternal and fetal fluids. In: *Fetal Physiology and Medicine.* Eds.: Beard, R.W. and Nathanielsz, P.W. (W.B. Saunders).
39. Barnes, R.J., Comline, R.S. and Silver, M. (1975) The effects of adrenalectomy and hypophysectomy in the foetal lamb. *J. Physiol.* 254, 15-16P.
40. Barraclough, C.H. and Cross, B.A. (1968) Unit activity on the hypothalamus of the cyclic female rat: effect of genital stimuli and progesterone. *J. Endocrinol.* 26, 339-359.
41. Bassett, J.M. (1974) Diurnal patterns of plasma insulin, growth hormone, corticosteroid and metabolite concentrations in fed and fasted sheep. *Aust. J. Biol. Sci.* 27, 167-181.
42. Bassett, J.M. and Hinks, N.T. (1969) Micro-determination of corticosteroids in ovine peripheral plasma: effects of venipuncture, corticotrophin, insulin and glucose. *J. Endocrinol.* 44, 387-403.
43. Bassett, J.M. and Madill, D. (1974) Influence of prolonged glucose infusions on plasma insulin and growth hormone concentrations of foetal lambs. *J. Endocrinol.* 62, 299-309.

44. Bassett, J.M., Mills, S.C. and Reid, R.L. (1966) The influence of cortisol on glucose utilization in sheep. *Metabolism* 15, 922-932.
45. Bassett, J.M., Oxborrow, T.J., Smith, I.D. and Thorburn, G.D. (1969) The concentration of progesterone in the peripheral plasma of the pregnant ewe. *J. Endocrinol.* 45, 449-457.
46. Bassett, J.M. and Thorburn, G.D. (1969) Foetal plasma corticosteroids and the initiation of parturition in sheep. *J. Endocrinol.* 44, 285-286.
47. Bassett, J.M. and Thorburn, G.D. (1973) Circulating levels of progesterone and corticosteroids in the pregnant ewe and its foetus. In: *The Endocrinology of Pregnancy and Parturition*. Ed.: Pierrepoint, C.G. (Tenovus Library Series, Alpha Omega Alpha, Cardiff), pp. 126-139.
48. Bassett, J.M., Thorburn, G.D. and Nichol, D.H. (1973) Regulation of insulin secretion in the ovine foetus in utero. Effects of sodium valerate on secretion and of adrenocorticotrophic-induced premature parturition on the insulin secretory response to glucose. *J. Endocrinol.* 56, 13-25.
49. Bassett, J.M., Thorburn, G.D. and Wallace, A.L.C. (1970) The plasma growth hormone concentration of the foetal lamb. *J. Endocrinol.* 48, 251-263.
50. Basson, P.A., Morgenthal, J.C., Bilbrough, R.B., Marais, J.L., Kruger, S.P. and Van der Merwe, J.L. de B. (1969) 'Grootlamsickte', a specific syndrome of prolonged gestation in sheep caused by a shrub, Salsola tuberculata (Fenzl ex Moq) schinz. var. Tomentosa C.A. Smith ex Aellen. *Onderstepoort J. Vet. Res.* 36 (1) 59-104.
51. Battaglia, F.C. and Meschia, G. (1973) Foetal metabolism and substrate utilization. In: *Foetal and Neonatal Physiology*. Eds.: Comline, R.S., Cross, K.W., Dawes, G.S. and Nathanielsz, P.W., pp. 382-397.
52. Bayard, F., Louveto, P., Ruckebusch, Y. and Boulard, U. (1972) Transplacental passage of dexamethasone in sheep. *J. Endocrinol.* 54, 349-350.
52a Bayard, F., Ances, I.G., Tapper, A.J., Weldon, V.V., Kowarski, A. and Migeon, C.J. (1970) Transplacental passage and fetal secretion of aldosterone. *J. Clin. Invest.* 49, 1389-1393.
53. Beamer, N., Hagemas, F. and Kittinger, G.W. (1972) Protein binding of cortisol in the rhesus monkey (*Macaca mulatta*). *Endocrinology* 90, 325-327.
54. Beck, J.S., Gordon, R.L., Donald, D., Melvin, J.M.O. (1969) Characterisation of antisera to a growth hormone like placental antigen (human placental lactogen). Immunofluorescence studies with these sera on normal and pathological syncyctiotrophoblast. *J. Pathol.* 97, 545-555.
55. Becks, H., Scow, R.C., Simpson, M.E., Asling, C.W., Li, C.II. and Evans, H.M. (1950) Response by the rat thyro-parathyroidectomised at birth to growth hormone and to thyroxine given separately or in combination. *Anat. Rec.* 107, 299-317.
56. Bedford, C.A., Challis, J.R.G., Harnson, F.A. and Heap, R.B. (1972) The role of oestrogens and progesterone in the onset of parturition in various species. *J. Reprod. Fertil.* suppl. 16, 1-23.
57. Bedford, C.A., Harrison, F.A. and Heap, R.B. (1972) The metabolic chearance rate and production rate of progesterone and the conversion to 20α-hydroxypregn-4-en-3 one in the sheep. *J. Endocrinol.* 55, 105-118.
58. Bedford, C.A., Harrison, F.A. and Heap, R.B. (1973) The kinetics of progesterone metabolism in the pregnant sheep. In: *The Endocrinology of Pregnancy and Parturi-*

tion. Ed.: Pierrepoint, C.G. (Tenovus Library Series, Alpha Omega Alpha, Cardiff) pp. 83-93.
59. Bedwani, J.R. and Marley, P.B. (1975) Enhanced inactivation of prostaglandin E_2 by the rabbit lung during pregnancy on progesterone treatment. *Br. J. Pharmacol.* 53, 547-554.
60. Beitins, I.Z., Bayard, F., Ances, I.G., Kowarski, A. and Migeon, C.J. (1973) Metabolic clearance rate, blood production, interconversion and transplacental passage of cortisol and cortisone in pregnancy near term. *J. Pediatr. Res.* 7, 509-519.
61. Beitins, I.Z., Kowarski, A., Shermeta, D.W., De Lemos, R.A. and Migeon, C.J. (1970) Fetal and maternal secretion rate of cortisol in sheep: Diffusion resistance of the placenta. *Pediatr. Res.* 4, 129-134.
62. Benirschke, K. (1956) Adrenals in anencephaly and hydrocephaly. *Obstet. Gynecol.* 8, 412-425.
63. Benirschke, K. and McKay, D.B. (1953) The antidiuretic hormone in fetus and infant. *Obstet. Gynecol.* 1, 638-649.
64. Berson, S.A. and Yalow, R.S. (1968) Radioimmunoassay of ACTH in plasma. *J. Clin. Invest.* 47, 2725-2751.
65. Best, M.M., Duncan, C.G. and Best, M.M. (1969) Accelerated maturation and persistent growth impairment in the rat resulting from thyroxine administration in the neonatal period. *J. Lab. Clin. Med.* 73, 135-143.
66. Bhalla, R. and Korenman, S.C. (1973) In discussion of paper by Liggins, G.C., Fairclough, R.J., Grieves, S.A., Kendall, J.Z. and Knox, B.S. *Rec. Prog. Horm. Res.* 29, 111-150.
67. Binns, W., Anderson, W.A. and Sullivan, D.J. (1960) Further observations on a congenital cyclopian-type malformation in lambs. *J. Am. Vet. Med. Assoc.* 137, 515-521.
68. Binns, W., James, J.F. and Shupe, J.L. (1964) Toxicosis of Veratrum californicum in ewes and its relationship to a congenital deformity in lambs. *Ann. NY Acad. Sci.* 111, 571-576.
69. Binns, W., Shupe, J.L., Keeler, R.F. and James, L.F. (1965) Chronologic evaluation of teratogenicity in sheep fed Veratrum californicum. *J. Am. Vet. Med. Assoc.* 147, 839-842.
70. Birnie, J.H. and Mapp, F.E. (1961) Thyroid function in foetal rats. *Fed. Proc.* 20, 201 (abstract).
71. Bisset, G.W., Clarke, B.J. and Haldar, J. (1970) Blood levels of oxytocin and vasopressin during suckling in the rabbit and the problem of their independent release. *J. Physiol.* 286, 711-722.
72. Blackburn, W.R. (1973) Hormonal influences in fetal lung development. In: *Respiratory Distress Syndrome*. Eds.: Villee, C.A., Villee, D.B. and Zuckerman, J. (Academic Press, New York), p. 271.
73. Blackburn, W.R., Travers, H. and Potter, D.M. (1972) The role of the pituitary–adrenal–thyroid axes in lung differentiation I. Studies of the cytology and physical properties of anencephalic rat lung. *Lab. Invest.* 26, 306-318.
74. Blair-West, J.R., Coghlan, J.P., Denton, D.A., Goding, J.R., Wintour, M. and Wright, R.D. (1963) *Rec. Prog. Horm. Res.* 19, 311-363.
75. Blatchley, F.R., Donovan, B.T., Poyser, N.L., Horton, E.W., Thompson, C.J. and Los, M. (1971) Identification of prostaglandin $F_{2\alpha}$ in the utero-ovarian blood of guinea-pig after treatment with oestrogen. *Nature* 230, 243-244.

76. Boddy, K. and Dawes, G.S. (1975) Fetal breathing. *Br. Med. Bull.* 31 (1) 3-7.
77. Boddy, K., Dawes, G.S., Fisher, R., Pinter, S. and Robinson, J.S. (1974) Foetal respiratory movements, electrocortical and cardiovascular responses to hypoxaemia and hypercapnia in sheep. *J. Physiol.* 243, 599-618.
78. Boddy, K., Dawes, G.S. and Robinson, J.S. (1973) A 24-hour rhythm in the foetus. In: *Foetal and Neonatal Physiology*. Proc. Sir Joseph Barcroft Centenary Symp. Eds.: Comline, R.S., Cross, K.W., Dawes, G.S. and Nathanielsz, P.W. (Cambridge University Press) pp. 63-66.
79. Boddy, K., Jones, C.T. and Robinson, J.S. (1974) Correlations between plasma ACTH concentrations and breathing movements in foetal sheep. *Nature* 250, 75-76.
80. Boddy, K., Jones, C.T., Mantell, C., Ratcliffe, J.G. and Robinson, J.S. (1974) Changes in plasma ACTH and corticosteroid of the maternal and fetal sheep during hypoxia. *Endocrinology* 94, 588-591.
81. Bosc, M.J. (1972) The induction and synchronization of lambing with the aid of dexamethasone. *J. Reprod. Fertil.* 28, 347-357.
82. Bosc, M.J. (1972) Conséquences sur la parturition de l'hypophysectomie de la mère ou du foetus chez la Brebis traitée par la dexamethasone. *C R Acad. Sci., Paris* 274, 93-96.
83. Bosc, M.J. and Fevre, J. (1974) Etude du mode d'action de la dexaméthasone utilisée pour induire l'agnelage chez la brebis. *C R Acad. Sci., Paris* 278, 315-318.
84. Bosu, W.T.K., Johansson, E.D.B. and Gemzell, C. (1974) Influence of oophorectomy, luteectomy, foetal death and dexamethasone on the peripheral plasma levels of oestrogens and progesterone in the pregnant macaca mulatta. *Acta Endocrinol.* 75, 601-616.
85. Brant, H.A. and Lackelin, G.C.L. (1971) Vibration of the cervix to expedite first stage of labour. *Lancet* 11, 686-687.
86. Bourne, G. (1970) The membranes. In: *Scientific Foundations of Obstetrics and Gynaecology*. Eds.: Philipp, E.E., Barnes, J. and Newton, M. (William Heinemann, London) pp. 181-185.
87. Brasel, J.A. and Winick, M. (1970) Differential cellular growth in the organs of hypothyroid rats. *Growth* 34, 197-207.
88. Brown, J.J., Davies, D.L., Doak, P.B., Lever, A.F., Robertson, J.I. and Tree, M. (1964) The presence of renin in the amniotic fluid. *Lancet* II, 64.
89. Brown-Grant, K. (1973) Recent studies on the sexual differentiation of the brain. In: *Foetal and Neonatal Physiology*. Proc. Sir Joseph Barcroft Centenary Symp. Eds.: Comline, R.S., Cross, K.W., Dawes, G.D. and Nathanielsz, P.W. (Cambridge University Press) pp. 527-545.
90. Brown-Grant, K. and Sherwood, M.R. (1971) The 'early androgen syndrome' in the guinea-pig. *J. Endocrinol.* 49, 277-291.
91. Broughton Pipkin, F. and Kirkpatrick, S.M.L. (1973) The blood volumes of foetal and newborn sheep. *Quart. J. Exp. Physiol.* 58, 181-188.
92. Broughton Pipkin, F., Lumbers, E.R. and Mott, J.C. (1974) Factors influencing plasma renin and angiotensin II in the conscious pregnant ewe and its fetus. *J. Physiol.* 243, 619-636.
93. Broughton Pipkin, F., Mott, J.C. and Roberton, N.R.C. (1971) Angiotensin II-like activity in circulating arterial blood in immature and adult rabbits. *J. Physiol.* 218, 385-403.

94. Bruce, N.W. and Norman, N. (1975) Influence of sexual dimorphism on foetal and placental weights in the rat. *Nature* 257, 62-63.
95. Bruchovsky, N. and Wilson, J.D. (1968) The intranuclear binding of testosterone and 5α-androstan-17β-ol-3-one by rat prostate. *J. Biol. Chem.* 243, 5953-5960.
96. Bruchovsky, N. and Wilson, J.D. (1968) The conversion of testosterone to 5α-androstan-17β-ol-3-one by rat prostate in vivo and in vitro. *J. Biol. Chem.* 243, 2012-2021.
97. Buckle, J.W. and Nathanielsz, P.W. (1975) Modification of myometrial activity in vivo by administration of cyclic nucleotides and theophylline to the pregnant rat. *J. Endocrinol.* 66, 339-347.
98. Buddingh, F., Parker, H.R., Ishizaki, G. and Tyler, W.S. (1971) Long term studies of the functional development of the fetal kidney in sheep. *Am. J. Vet. Res.* 32, 1993-1998.
99. Burrow, G.N. and Anderson, G.G. (1970) Bidirectional transfer of thyroxine in sheep. *Clin. Res.* 18, 356.
100. Burton, A.M. and Forsling, M.L. (1972) Hormone content of the neurohypophysis in fetal, newborn and adult guinea-pigs. *J. Physiol.* 221, 6-7P.
100a Burton, A.M., Illingworth, D.V., Challis, J.R.G. and McNeilly, A.S. (1974) Placental transfer of oxytocin in the guinea-pig and its release during parturition. *J. Endocrinol.* 60, 499-506.
101. Campbell, A.L., Leong, M.K. and Murphy, B.E.P. (1975) Physiologically active cortisol (PAF) in mother and fetus at human parturition. *Gynecol. Invest.*
102. Campbell, H.J. (1966) The development of the primary portal plexus in the median eminence of the rabbit. *J. Anat.* 100, 381-387.
103. Campbell, S., Wladimiroff, J.W. and Dewhurst, C.J. (1973) The antenatal measurement of fetal urine production. *J. Obstet. Gynecol. Br. Commonw.* 80, 680-686.
104. Carr, E.A., Beierwaltes, W.H., Raman, G., Dodson, V.N., Tanton, J., Betts, J.S. and Stambaugh, R.A. (1959) The effect of maternal thyroid function on foetal thyroid function and development. *J. Clin. Endocrinol. Metab.* 19, 1-18.
105. Carretero, O.A., Bujak, B., Hodari, A.A., Hodgkinson, C.P. and Bumpus, F.M. (1971) Identification of a pressor polypeptide in human amniotic fluid. *Am. J. Obstet. Gynecol.* III, 1075-1082.
106. Cawson, M.J., Anderson, A.B.M., Turnbull, A.C. and Lampe, L. (1974) Cortisol, cortisone and 11-deoxycortisol levels in human umbilical and maternal plasma in relation to the onset of labour. *J. Obstet. Gynaecol. Br. Commonw.* 81, 737-745.
107. Challis, J.R.G. (1971) Sharp increase in free circulating oestrogens immediately before parturition in sheep. *Nature* 229, 208.
108. Challis, J.R.G. (1974) Physiology and pharmacology of prostaglandins in parturition. In: *Prostaglandins. Population Report*, No. 5, G.45-G.53.
109. Challis, J.R.G., Davies, I.J., Benirschke, K., Hendrickx, A.G. and Ryan, K.J. (1974) The concentrations of progesterone, estrone and estradiol 17-β in the peripheral plasma of the rhesus monkey during the final third of gestation and after the induction of abortion with $PGF_{2\alpha}$. *Endocrinology* 95, 547-553.
110. Challis, J.R.G., Davies, I.J., Benirschke, K., Hendrickx, A.G. and Ryan, K.J. (1974) The effects of dexamethasone on plasma steroid levels and fetal adrenal histology in the pregnant rhesus monkey. *Endocrinology* 95, 1300-1305.
111. Challis, J.R.G., Davies, I.J., Benirschke, K., Hendrickx, A.G. and Ryan, K.J. (1975) The effects of dexamethasone on the peripheral plasma concentrations of

androstenedione, testosterone and cortisol in the pregnant rhesus monkey. *Endocrinology* 96, 185-192.
112. Challis, J.R.G., Harrison, F.A. and Heap, R.B. (1971) Uterine production of oestrogens and progesterone at parturition in sheep. *J. Reprod. Fertil.* 12, 369-372.
113. Challis, J.R.G., Harrison, F.A., Heap, R.B., Horton, E.W. and Poyser, N.L. (1972) A possible role of oestrogens in the stimulation of prostaglandin $F_{2\alpha}$ output at the time of parturition in a sheep. *J. Reprod. Fertil.* 30, 485-488.
114. Challis, J.R.G., Osathanondh, R., Ryan, K.J. and Tulchinsky, D. (1974) Maternal and fetal plasma prostaglandin levels at vaginal delivery and caesarian section. *Prostaglandins* 6, 281-288.
114a Challis, J.R.G., Robinson, J.S. and Thorburn, G.D. (1977) Fetal and maternal endocrine changes during pregnancy and parturition in the rhesus monkey. In: *The Fetus and Birth*. Ed.: Maeve O'Connor, CIBA Foundation Symposium, No. 47, New Series. (Elsevier, Amsterdam).
115. Challis, J.R.G. and Thorburn, G.D. (1975) Prenatal endocrine function and the initiation of parturition. In: *Perinatal Research*. Ed.: Nathanielsz, P.W., *Br. Med. Bull.* 31 (1) 57-61.
116. Challis, J.R.G. and Thorburn, G.D. (1976) The fetal pituitary–adrenal axis and its functional interactions with the neurohypophysis. In: *Fetal Physiology and Medicine*, Eds.: Beard, R.W. and Nathanielsz, P.W. (Saunders).
117. Chapman, R.E., Hopkins, P.S. and Thorburn, G.D. (1974) The effects of foetal thyroidectomy and thyroxine administration on the development of the skin and wool follicles of sheep foetuses. *J. Anat.* 117, 419-432.
118. Chard, T. (1972) The posterior pituitary in human and animal parturition. *J. Reprod. Fertil.* suppl. 16, 121-128.
119. Chard, T., Boyd, N.R.H., Forsling, M.L., McNeilly, A.S. and Landon, J. (1970) The development of a radioimmunoassay for oxytocin: the extraction of oxytocin from plasma and the measurement during parturition in human and goat blood. *J. Endocrinol.* 48, 223-234.
120. Chard, T., Hudson, C.N., Edwards, C.R.W. and Boyd, N.R.H. (1971) Release of oxytocin and vasopressin by the human foetus during labour. *Nature* 234, 352-354.
120a Chayen, J., Loveridge, N. and Daly, J.R. (1972) A sensitive bioassay for adrenocorticotrophic hormone in human plasma. *Clin. Endocrinol.* 1, 219-233.
121. Chester Jones, I., Jarrett, I.G., Vinson, G.P. and Potter, K. (1964) Adrenocorticosteroid production of foetal sheep near term. *J. Endocrinol.* 29, 211-212.
122. Chez, R.A., Hutchinson, D.L., Salazar, H. and Mintz, D.H. (1970) Some effects of fetal and maternal hypophysectomy in pregnancy. *Am. J. Obstet. Gynecol.* 108, 643-650.
123. Chez, R.A., Mintz, D.H., Horger, E.O. and Hutchinson, D.L. (1970) Factors affecting the response to insulin in the normal subhuman pregnant primate. *J. Clin. Invest.* 49, 1517-1527.
124. Chiswick, M.L., Ahmed, A., Jack, P.M.B. and Milner, R.D.G. (1973) Control of fetal lung development in the rabbit. *Arch. Dis. Child.* 48, 709-713.
125. Chopra, I.J., Sack, J. and Fisher, D.A. (1975) Circulating 3,3′,5′-triiodothyronine (reverse T3) in the human newborn. *J. Clin. Invest.* 55, 1137-1141.
126. Chopra, I.J., Sack, J. and Fisher, D.A. (1975) 3,3′,5′-Triiodothyronine (reverse T3) and 3,3′,5-Triiodothyronine (T3) in fetal and adult sheep: studies of metabolic

clearance rate, production rate, serum binding and thyroidal content relative to thyroxine. *Endocrinology* 97, 1080-1088.
127. Christopherson, R.J. and Webster, A.J.F. (1972) Changes during eating in oxygen consumption, cardiac function and body fluids of sheep. *J. Physiol.* 221, 441-457.
128. Clayton, G.W., Libritz, L., Gardner, R.L. and Guillemin, R. (1963) Studies on the circadian rhythm of pituitary adrenocorticotrophin release in man. *J. Clin. Endocrinol. Metab.* 23, 975-980.
129. Clements, J.A. and King, R.J. (1973) Pulmonary surfactant and its assay. In: *Foetal and Neonatal Physiology.* Proc. Sir Joseph Barcroft Centenary Symp. Eds.: Comline, R.S., Cross, K.W., Dawes, G.D. and Nathanielsz, P.W. (Cambridge University Press) pp. 618-622.
130. Cohen, A. (1973) Plasma corticosterone concentration in the fetal rat. *Horm. Metab. Res.* 5, 66.
131. Cohn, H.E., Sacks, E.J., Heyman, M.A. and Rudolph, A.M. (1974) Cardiovascular responses to hypoxaemia and acidemia in fetal lambs. *Am. J. Obstet. Gynecol.* 120, 817-824.
132. Colby, H.D., Malendowicz, L.K., Caffrey, J.L. and Kitay, J.I. (1974) Effects of hypophysectomy and ACTH on adrenocortical function in the rat. *Endocrinology* 94, 1346-1350.
133. Comerford, J.B. (1965) Pregnancy with anencephaly. *Lancet* 1, 679-680.
134. Comline, R.S., Nathanielsz, P.W. and Silver, M. (1970) Passage of thyroxine across the placenta in the foetal sheep. *J. Physiol.* 207, 3-4P.
135. Comline, R.S., Nathanielsz, P.W., Paisey, R.B. and Silver, M. (1970) Cortisol turnover in the sheep foetus immediately prior to parturition. *J. Physiol.* 210, 141-142P.
136. Comline, R.S. and Silver, M. (1965) Development of activity in the adrenal medulla of the fetus and newborn animal. *Br. Med. Bull.* 22, 16-20.
137. Comline, R.S. and Silver, M. (1970) Daily changes in foetal and maternal blood of conscious pregnant ewes with catheters in umbilical and uterine vessels. *J. Physiol.* 209, 567-586.
138. Comline, R.S. and Silver, M. (1972) The composition of foetal and maternal blood during parturition in the ewe. *J. Physiol.* 222, 233-256.
139. Comline, R.S. and Silver, M. (1974) Recent observations on the undisturbed fetus in 'utero' and its delivery. In: *Recent Advances in Physiology*, No. 9. Ed.: Linden, R.J. (Churchill Livingstone, Edinburgh) pp. 406-454.
140. Comline, R.S. and Silver, M. (1975) Placental transfer of blood gases. *Br. Med. Bull.* 35 (1). Ed.: Nathanielsz, P.W. pp. 25-31.
141. Conklin, P.M., Schindler, W.J. and Hull, S.F. (1973) Hypothalamic thyrotrophin releasing factor: activity and pituitary responsiveness during development in the rat. *Neuroendocrinology* 11, 197-211.
142. Coslovsky, R. and Yalow, R.S. (1974) Influence of the hormonal forms of ACTH on the pattern of corticosteroid secretion. *Biochem. Biophys. Res. Commun.* 60, 1351-1356.
143. Craft, I.L., Scrivener, R. and Dewhurst, C.J. (1973) Prostaglandin $F_{2\alpha}$ levels in the maternal and fetal circulations in late pregnancy. *J. Obstet. Gynaecol. Br. Commonw.* 80(7) 616-618.
144. Cramer, D.W., Beck, P. and Makowski, E.L. (1971) Correlation of gestational age with maternal human chorionic somatomammotrophin and maternal and

fetal growth hormone plasma concentrations during labour. *Am. J. Obstet. Gynecol.* 109, 649-655.
145. Crawford, J.D. and McCance, R.A. (1960) Sodium transport by the chorioallantoic membrane of the pig. *J. Physiol.* 151, 458-471.
146. Csapo, A. (1955) The mechanism of myometrial function and its disorders. *Mod. Trends Obstet. Gynaecol.* 2, 20-49.
147. Csapo, A. (1956) The mechanism of effect of the ovarian steroids. *Rec. Prog. Horm. Res.* 12, 405-431.
148. Csapo, A.I. and Csapo, E.E. (1974) The 'prostaglandin step', a bottle-neck in the activation of the uterus. *Life Sci.* 14, 719-724.
149. Csapo, A.I., Knobil, E., Van der Molen, H.J. and Wiest, W.G. (1971) Peripheral plasma progesterone levels during human pregnancy and labour. *Am. J. Obstet. Gynecol.* 110, 630-632.
150. Csapo, A.I., Pohanka, O. and Kaihola, H.L. (1974) Progesterone deficiency and premature labour. *Br. Med. J.* 1, 137-140.
151. Currie, W.B., Wong, M.S.F., Cox, R.I. and Thorburn, G.D. (1973) Spontaneous or dexamethasone-induced parturition in the sheep and goat: changes in plasma concentrations of maternal prostaglandin F and foetal oestrogen sulphate. *Mem. Soc. Endocrinol.* 20, 95-118.
152. Dale, H.M. (1956) On some physiological actions of ergot. *J. Physiol.* 34, 163–206.
153. Daniel, P.M. and Prichard, M.M.L. (1957) The vascular arrangements of the pituitary gland of the sheep. *Quart. J. Exp. Physiol.* 42, 237-248.
154. D'Angelo, S.A. (1967) Pituitary and thyroid interrelationships in maternal, foetal and neonatal guinea pigs. *Endocrinology* 81, 132-138.
155. D'Angelo, S.A. and Wall, N.R. (1971) Simultaneous effects of thyroid and adrenal inhibitors on maternal-fetal endocrine interrelations in the rat. *Endocrinology* 89, 591-597.
156. D'Angelo, S.A. and Wall, N.R. (1972) Maternal–fetal endocrine interrelations: effects of synthetic thyrotrophin releasing hormone (TRH) on the fetal pituitary–thyroid system of the rat. *Neuroendocrinology* 9, 197-206.
157. Davies, I.J., Challis, J.R.G. and Ryan, K.J. (1974) Progesterone receptors in the myometrium of pregnant rabbits. *Endocrinology* 95, 165-173.
158. Davies, I.J. and Ryan, K.J. (1972) Comparative endocrinology of gestation. *Vitam. Horm.* 30, 223-278.
159. Davies, I.J. and Ryan, K.J. (1973) Glucocorticoid synthesis from pregnenolone by sheep foetal adrenals in vitro. *J. Endocrinol.* 58, 485-491.
160. Davis, S.L. (1972) Plasma levels of prolactin, growth hormone and insulin in sheep following the infusion of arginine, leucine and phenylalanine. *Endocrinology* 91, 549-555.
161. Davis, S.L. and Borger, M.L. (1973) Metabolic clearance rates and secretion rates of prolactin in sheep. *Endocrinology* 92, 1414-1418.
162. Dawes, G.S. (1968) *Foetal and Neonatal Physiology.* (Yearbook Medical Publishers, Chicago.)
163. Dawes, G.S. (1973) Breathing and rapid eye movement sleep before birth. In: *Foetal and Neonatal Physiology.* Proc. Sir Joseph Barcroft Centenary Symp. Eds.: Comline, R.S., Cross, K.W., Dawes, G.S. and Nathanielsz, P.W. (Cambridge University Press) pp. 49-62.
164. Dawes, G.S., Duncan, S.L.B., Lewis, B.V., Merlet, C.L., Owen-Thomas, J.B.

and Reeves, J.T. (1969) Hypoxaemia and aortic chemoreceptor function in foetal lambs. *J. Physiol.* 201, 105-116.
165. Dawes, G.S., Fox, H.E., Leduc, B.M., Liggins, G.C. and Richards, R.T. (1972) Respiratory movements and rapid eye movement sleep in the foetal lamb. *J. Physiol.* 220, 119-143.
166. Daughaday, W.H. and Jacobs, L.S. (1972) Human prolactin. *Ergeb. Physiol.* 67, 169-194.
167. Dazard, A., Gallet, D. and Saez, J.M. (1975) Adenyl cyclase activity in rat, ovine and human adrenal preparations. *Horm. Metab. Res.* 7, 184-189.
168. Denamur, R. and Martinet, J. (1955) Effects de l'ovariectomie chez la brebis pendant la gestation. *CR Hebd. séances Mém. Biol.* 149, 2105-2107.
169. Denamur, R. and Martinet, J. (1961) Effets de l'hypophysectomie et de la section de la tige pituitaire sur la gestation de la brebis. *Ann. Endocrinol.* 21, 755-759.
170. Dewar, A.D. (1968) Litter size and the duration of pregnancy in mice. *Quart. J. Exp. Physiol.* 53, 155-161.
171. Diczfalusy, E. and Mancuso, S. (1969) Oestrogen metabolism in pregnancy. In: *Foetus and Placenta*. Eds.: Klopper, A. and Diczfalusy, E. (Blackwell, Oxford) pp. 191-248.
172. Dixon, R., Hyman, A., Gurpide, E., Dyrenfurth, I., Cohen, H., Bowe, E., Engel, T., Daniel, S., James, S. and Van de Wiele, R. (1970) Feto–maternal transfer and production of cortisol in the sheep. *Steroids* 16, 771-789.
173. Dostal, M. and Jelinck, R. (1971) Introduction of cleft palate in rats with intraamniotic corticoids. *Nature* 230, 464-465.
174. Drost, M. and Holm, L.W. (1968) Prolonged gestation in ewes after fetal adrenalectomy. *J. Endocrinol.* 40, 293-296.
175. Drost, M., Kumagai, L.F. and Guzman, M. (1973) Sequential foetal-maternal plasma cortisol levels in sheep. *J. Endocrinol.* 56, 483-492.
176. Dubois, P.M., Paulin, C., Assan, R. and Dubois, M.P. (1975) Evidence for immunoreactive somatostatin in the endocrine cells of human foetal pancreas. *Nature* 256, 731-732.
177. Dunham, E.W., Haddox, M.K. and Goldberg, N.D. (1974) Alteration of vein cyclic 3′,5′ nucleotide concentrations during changes in contractility. *Proc. Nat. Acad. Sci. U.S.A.* 71, 815-819.
178. Dupouy, J.P. (1974) Site of the negative feedback action of corticosteroids on the hypothalamo-hypophysial system of the rat fetus. *Neuroendocrinology* 16, 148-155.
179. Dupouy, J.P., Coffigny, H. and Magre, S. (1975) Maternal and foetal corticosterone levels during late pregnancy in rats. *J. Endocrinol.* 65, 347-352.
180. Dupouy, J.P. and Cohen, A. (1975) Comparaison de l'activité corticosurrénalienne foetale et maternelle au cours du nycthémère et durant la gestation. *CR Acad. Sci. Paris* 280, 463-466.
181. Dussault, J.H., Hobel, C.J., Distefano, J.J., Erenberg, A. and Fisher, D.A. (1972) Triiodothyronine turnover in maternal and fetal sheep. *Endocrinology* 90, 1307-1308.
182. Dussault, J., Vas V. Row, Lickrish, G. and Volpe, R. (1969) Studies of serum triiodothyronine concentration in maternal and cord blood: transfer of T3 across the human placenta. *J. Clin. Endocrinol. Metab.* 29, 595-603.
183. Du Vigneaud, V., Ressler, C. and Trippett, S. (1953) The sequence of amino-

acids in oxytocin with a proposal for the structure of oxytocin. *J. Biol. Chem.* 205, 949-957.

184. Eayrs, J.T. (1954) The vascularity of the cerebral cortex in normal and cretinous rats. *J. Anat.* 88, 164-173.
185. Eayrs, J.T. (1961) Age as a factor determining the severity and reversibility of the effects of thyroid deprivation in the rat. *J. Endocrinol.* 22, 409-419.
186. Eayrs, J.T. and Holmes, R.L. (1964) Effect of neonatal hyperthyroidism on pituitary structure and function in the rat. *J. Endocrinol.* 29, 71-81.
187. Edelman, I.S. and Ismail-Beigi, F. (1974) Thyroid thermogenesis and active sodium transport. *Rec. Prog. Horm. Res.* 30, 235-254.
188. Edwards, C.R.W., Chard, T., Kitau, M.J. and Forsling, M.L. (1970) The development of a radioimmunoassay and a plasma extraction method for vasopressin. *J. Endocrinol.* 48, xi-xii.
189. Eguchi, Y., Eguchi, K. and Wells, L.J. (1964) Compensatory hypertrophy of right adrenal after left adrenalectomy: observations in fetal, newborn and week-old rats. *Proc. Soc. Exp. Biol. Med.* 116, 89-92.
190. Ellis, S.T., Beck, J.S. and Currie, A.R. (1966) Cellular localisation of Growth Hormone in the human foetal adenohypophysis. *J. Pathol. Bacteriol.* 92, 179-183.
191. Elliot, R.B. and Starling, M.B. (1972) The effect of $PGF_{2\alpha}$ in the closure of the ductus arteriosus. *Prostaglandins* 2, 399-403.
192. Embrey, M.P. (1971) Prostaglandins. *Proc. R. Soc. Med.* 64, 1018-1020.
193. Erenberg, A. and Fisher, D.A. (1973) Thyroid hormone metabolism in the foetus. In: *Foetal and Neonatal Physiology*. Proc. Sir Joseph Barcroft Centenary Symp. Eds.: Comline, R.S., Cross, K.W., Dawes, G.D. and Nathanielsz, P.W. (Cambridge University Press) pp. 508-526.
194. Erenberg, A., Omor, K., Menkes, J.H., Oh, W. and Fisher, D.A. (1974) Growth and development of the thyroidectomized ovine fetus. *Pediatr. Res.* 8, 783.
195. Escobar del Rey, F. and Morreale de Escobar, G. (1961) The effect of propylthiouracil, methylthiouracil and thiouracil on the peripheral metabolism of l-thyroxine in thyroidectomised, l-thyroxine maintained rats. *Endocrinology* 69, 456-465.
196. Fairclough, R.J. and Liggins, G.C. (1975) Protein binding of plasma cortisol in the foetal lamb near term. *J. Endocrinology* 67, 555-564.
197. Fajer, A.B., Holzbauer, M. and Newport, H.M. (1971) The contribution of the adrenal gland to the total amount of progesterone produced in the female rat. *J. Physiol.* 214, 115-126.
198. Falin, L. (1961) The development of human hypophysis and differentiation of its anterior lobe during embryonic life. *Acta Anat.* 44, 188-205.
199. Farrell, P.M. (1973) Regulation of pulmonary lecithin synthesis. In: *Respiratory Distress Syndrome*. Eds.: Villee, C.A., Villee, D.B. and Zuckerman, J. (Academic Press, New York) pp. 311-341.
200. Farrell, P.M. and Zachman, R.D. (1973) Induction of choline phosphotransferase and lecithin synthesis in the fetal lung by corticosteroids. *Science* 179, 297-298.
201. Feder, H.H. (1971) The comparative actions of testosterone proprionate and 5α-androstan-17β-ol-3-one proprionate on the reproductive behaviour, physiology and morphology of male rats. *J. Endocrinol.* 51, 241-252.
202. Feldman, J.D., Vazquez, J.J. and Kurtz, S.M. (1961) Maturation of the rat foetal thyroid. *J. Biophys. Biochem. Cytol.* 11, 365-383.

203. Findlay, J.K. and Cox, R.I. (1970) Oestrogens in the plasma of the sheep foetus. *J. Endocrinol.* 46, 281-282.
204. Findlay, J.K. and Seamark, R.F. (1973) The occurrence and metabolism of oestrogens in the sheep foetus and placenta. In: *The Endocrinology of Pregnancy and Parturition – Experimental Studies in the Sheep*. Ed.: Pierrepoint, C.G. (Tenovus Library Series, Alpha Omega Alpha, Cardiff), pp. 54-64.
205. Fink, G. and Smith, G.C. (1971) Ultrastructural features of the developing hypothalamo-hypophyseal axis in the rat. *Z. Zellforsch. Mikrosk. Anat.* 119, 208-226.
206. Fisch, L., Sala, N.L. and Schwarz, R.L. (1964) Effect of cervical dilatation upon uterine contractility in pregnant women and its relation to oxytocin secretion. *Am. J. Obstet. Gynecol.* 90, 108-114.
207. Fisher, D.A. and Dussault, J.H. (1971) Contribution of methodological artifacts to the measurement of T3 concentration in serum. *J. Clin. Endocrinol. Metab.* 32, 675-679.
208. Fisher, D.A. and Dussault, J.H. (1974) Development of the mammalian thyroid gland. In: *Handbook of Physiology*, Section 7. Endocrinology Vol. III. Thyroid. (Am. Physiol. Soc.) pp. 21-38.
209. Fisher, D.A., Dussault, J.H., Hobel, C.J. and Lam, R. (1973) Serum and thyroid gland triiodothyronine in the human fetus. *J. Clin. Endocrinol. Metabol.* 36, 397-400.
210. Fisher, D.A., Hobel, C.J., Garza, R. and Pierce, C.A. (1970) Thyroid function in the preterm fetus. *Pediatrics* 46, 208-215.
211. Fisher, D.A. and Oddie, T.H. (1964) Neonatal thyroidal hyperactivity – response to cooling. *Am. J. Dis. Child.* 107, 574-581.
212. Fisher, D.A. and Odell, W.D. (1969) Acute release of thyrotrophin in the newborn. *J. Clin. Invest.* 48, 1670-1677.
213. Fisher, D.A., Odell, W.D., Hobel, C.J. and Garza, R. (1969) Thyroid function in the term fetus. *Pediatrics* 44, 526-535.
214. Fisher, D.A., Lehman, H. and Lackey, C. (1964) Placental transport of thyroxine. *J. Clin. Endocrinol. Metab.* 24, 393-400.
215. Fitzpatrick, R.J. (1961) The estimation of small amounts of oxytocin in blood. In: *Oxytocin*. Eds.: Caldeyro-Bareia and Heller, H. (Pergamon Press, Oxford).
216. Fleischer, N. and Vale, W. (1968) Inhibition of vasopressin-induced ACTH release from the pituitary by glucocorticoids in vitro. *Endocrinology* 83, 1232-1236.
217. Flint, A.P., Anderson, A.B.M., Patten, P.T. and Turnbull, A.C. (1974) Control of utero-ovarian venous prostaglandin F during labour in the sheep: acute effects of vaginal and cervical stimulation. *J. Endocrinol.* 63, 67-87.
218. Flint, A.P.F., Anderson, A.B.M., Steele, P.A. and Turnbull, A.C. (1975) Effect of foetal adrenalectomy on maternal oestrogen levels at dexamethasone-induced parturition in sheep. *J. Endocrinol.* 67, 25p.
219. Flint, A.P.F., Goodson, J.D. and Turnbull, A.C. (1975) Increased levels of 17α,20α-dihydroxypregn-4-en-3-one in utero-ovarian venous plasma near parturition in pregnant sheep. *J. Endocrinol.* 65, 41P-42P.
220. Florsheim, W.H., Faircloth, M.A., Corcorran, N.L. and Rudko, P. (1966) Perinatal thyroid function in the rat. *Acta Endocrinol.* 52, 375-382.
221. Florsheim, W.H. and Rudko, L. (1968) The development of portal system function in the rat. *Neuroendocrinology* 3, 89-98.
222. Folley, S.J. and Knaggs, G.S. (1965) Levels of oxytocin in the jugular vein blood of goats during parturition. *J. Endocrinol.* 33, 301-315.

223. Forsling, M., Jack, P.M.B. and Nathanielsz, P.W. (1975) Lack of placental transport of oxytocin from foetal lamb to pregnant ewe between 114 and 134 days gestation. *J. Endocrinol.* 64, 41P.
224. Forsling, M., Jack, P.M.B. and Nathanielsz, P.W. (1975) Plasma oxytocin concentrations in the foetal sheep. *Horm. Metab. Res.* 7, 197-198.
225. Forsling, M.L., Jones, J.J. and Lee, J. (1968) Factors influencing the sensitivity of the rat to vasopressin. *J. Physiol.* 196, 495-505.
226. France, J.T. and Liggins, G.C. (1969) Placental sulphatase deficiency. *J. Clin. Endocrinol. Metab.* 29, 138-141.
227. Friesen, H. (1965) Purification of a placental factor with immunological and chemical similarity to human growth hormone. *Endocrinology* 76, 369-381.
228. Friesen, H. (1965) Further purification and characterisation of a placental protein with immunological similarity to human growth hormone. *Nature* 208, 1214-1215.
229. Friesen, H., Belanger, C., Gwyda, G. and Hwang, P. (1972) The synthesis and secretion of placental lactogen and pituitary prolactin. In: *Lactogenic Hormones*. Eds.: Wolstenholme, G.E.W. and Knight, J. (Churchill Livingstone, Edinburgh and London), pp. 83-103.
230. Fuchs, A.R., Coutinho, E.M., Xavier, R., Bates, P.E. and Fuchs, F. (1968) Effect of ethanol on the activity of the non-pregnant human uterus and its reactivity to neurohypophyseal hormones. *Am. J. Obstet. Gynecol.* 101, 997-1000.
231. Fujita, T., Eguchi, Y., Morikawa, Y. and Hashimoto, Y. (1970) Hypothalamic–hypophysial–adrenal and thyroid systems: observations in fetal rats subjected to hypothalamic destruction, brain compression and hypervitaminosis, A. *Anat. Rec.* 166, 659-672.
232. Fukuda, H., Yasuda, N., Greer, M.A., Kutas, M. and Greer, S.E. (1975) Changes in plasma thyroxine, triiodothyronine and TSH during adaptation to iodine deficiency in the rat. *Endocrinology* 97, 307-314.
233. Fylling, P. (1969) Serial progesterone assays in peripheral blood of sheep during the last week of pregnancy. *Acta Endocrinol.* suppl. 138, 251.
234. Fylling, P. (1970) The effect of pregnancy, ovariectomy and parturition on plasma progesterone levels in sheep. *Acta Endocrinol.* Kbh 65, 273-283.
235. Fylling, P. (1971) Premature parturition following dexamethasone administration to pregnant ewes. *Acta Endocrinol.* 66, 289-295.
236. Ganjam, V.K., Campbell, A.L. and Murphy, B.E.P. (1972) Changing patterns of circulating corticosteroids in rabbits following prolonged treatment with ACTH. *Endocrinology* 91, 607-611.
237. Ganong, W.F. and Hume, D.M. (1956) The effect of unilateral adrenalectomy of adrenal venous 17-hydroxycorticosteroid output in the dog. *Endocrinology* 59, 302-305.
238. Geloso, J.P. (1961) Date de l'entree en fonction de la thyroide chez le foetus de rat. *CR Hebd. Séances Mém. Soc. Biol.* 155, 1239-1244.
239. Geloso, J.P. and Bernard, G. (1967) Effets de l'ablation de la thyroide maternelle ou foetale sur le taux des hormones circulantes chez le foetus de rat. *Acta Endocrinol.* 56, 561-566.
240. Geloso, J.P., Hemon, P., Legrand, J., Legrand, C. and Jost, A. (1968) Some aspects of thyroid physiology during the perinatal period. *J. Gen. Comp. Endocrinol.* 10, 191-197.
241. Genazzani, A.R., Fraioli, F., Hurlimann, J., Fioretti, P. and Felber, J.P. (1975)

Immunoreactive ACTH and cortisol plasma levels during pregnancy. Detection and partial purification of corticotrophin-like placental hormone: The human chorionic corticotrophin (HCC). *J. Clin. Endocrinol.* 4, 1-14.
242. Giannopoulos, G., Mulay, S. and Solomon, S. (1972) Cortisol receptors in rabbit fetal lung. *Biochem. Biophys. Res. Commun.* 47, 411-418.
243. Gillespie, A., Brummer, H.C. and Chard, T. (1972) Oxytocin release by infused prostaglandin. *Br. Med. J.* 1, 543-544.
244. Glendening, M.B., Titus, M.A., Schroeder, S.A., Mohun, G. and Page, E.W. (1965) The destruction of oxytocin and vasopressin by the aminopeptidases in sera from pregnant women. *Am. J. Obstet. Gynaecol.* 92, 814-820.
245. Glydon, R.S. (1957) Development of the blood supply of the pituitary in the albino rat with special reference to the portal vessels. *J. Anat.* 91, 237-244.
246. Goldberg, N.D., Haddox, M.K., Hartle, D.K. and Haddon, J.W. (1973) Cellular mechanisms. In: *Fifth Int. Congr. Pharmacol., San Francisco 1972*. Eds.: Maxwell, R.A. and Acheson, G.H. (Karger, Basel) pp. 146-169.
247. Gorbman, A., Waterman, A., Barnes, C.M. and Bustad, L.K. (1957) Thyroidal function in foetal and pregnant sheep given chronic low level dosages of I^{131}. *Endocrinology* 60, 565-567.
248. Greenberg, A.H., Czernichow, P., Reba, R.C., Tyson, J. and Blizzard, R.N. (1970) Observations on the maturation of thyroid function in early fetal life. *J. Clin. Invest.* 49, 1790-1803.
249. Greenberg, A.H., Najjar, S. and Blizzard, R.M. (1974) Effects of thyroid hormone on growth, differentiation and development. In: *Handbook of Physiology*, Section 7. Endocrinology, Vol. III Thyroid (Am. Physiol. Soc.) pp. 377-387.
250. Greiss, F.G. and Miller, H.B. (1971) Unilateral control of uterine blood flow in the ewe. *Am. J. Obstet. Gynecol.* 111, 299-301.
251. Gresham, E.L., Rankin, J.H.G., Makowski, E.L., Meschia, G. and Battaglia, F.C. (1972) An evaluation of fetal renal function in a chronic sheep preparation. *J. Clin. Invest.* 51, 149-156.
252. Gross, F., Schaechtelin, G., Ziegler, M. and Berger, M. (1964) A renin-like substance in the placenta and uterus of the rabbit. *Lancet* 1, 914.
253. Grumbach, M.M. and Kaplan, S.L. (1973) Ontogenesis of growth hormone, insulin, prolactin and gonadotropin secretion in the human fetus. In: *Foetal and Neonatal Physiology*. Eds.: Comline, R.S., Cross, K.W., Dawes, G.D. and Nathanielsz, P.W. (Cambridge University Press) pp. 462-487.
254. Grumbach, M.M., Kaplan, S.L., Sciarra, J.J. and Burr, I.M. (1968) Chorionic growth hormone prolactin (CGP): secretion, disposition, biologic activity in man, and postulated function as the 'growth hormone' of the second half of pregnancy. *Ann. NY Acad. Sci.* 148, 501-531.
255. Grumbach, M.M. and Werner, S.C. (1956) Transfer of thyroid hormone across the human placenta at term. *J. Clin. Endocrinol. Metab.* 16, 1392-1395.
256. Grunt, J.A. and Reynolds, D.W. (1970) Insulin, blood sugar and growth hormone levels in an anencephalic infant before and after intravenous administration of glucose. *Pediatrics* 76, 112-116.
257. Gurpide, E., Tseng, J., Escareena, L., Fahning, M., Gibson, C. and Fehr, P. (1972) Fetomaternal production and transfer of progesterone and uridine in sheep. *Am. J. Obstet. Gynecol.* 113, 21-32.
258. Gustavii, B. (1972) Labour: A delayed menstruation? *Lancet* 2, 1149-1150.

259. Gustavii, B. and Brunk, U. (1972) A histological study of the effect of the placenta of intra-amniotically and extra-amniotically injected hypertonic saline in therapeutic abortion. *Acta Obstet. Gynecol. Scand.* 51, 121-125.
260. Gustavii, B. and Green, K. (1972) Release of prostaglandin $F_{2\alpha}$ following injection of hypertonic saline for therapeutic abortion: A preliminary study. *Am. J. Obstet. Gynecol.* 114, 1099-1100.
260a Hagemenas, F.C., Baughman, W.L. and Kittinger, G.W. (1975) The effect of fetal hypophysectomy on placental biosynthesis of progesterone in rhesus. *Endocrinology* 96, 1059-1062.
261. Hagemenas, F.C. and Kittinger, G.W. (1972) The influence of fetal sex on plasma progesterone levels. *Endocrinology* 91, 253-256.
262. Hagemenas, F.C. and Kittinger, G.W. (1973) The influence of fetal sex on the levels of plasma progesterone in the human fetus. *J. Clin. Endocrinol. Metab.* 36, 389-391.
263. Halasz, B., Kasaras, B. and Lengvari, I. (1972) Ontogenesis of the neurovascular link between the hypothalamus and the anterior pituitary in the rat. In: *Brain-Endocrine Interaction.* Eds.: Knigge, K.M., Scott, D.E. and Weindl, A.S. (Karger, Basel) p. 27.
264. Haldar, J. (1970) Independent release of oxytocin and vasopressin during parturition in the rabbit. *J. Physiol.* 206, 723-730.
265. Hales, J.R.S., Hopkins, P.S. and Thorburn, G.D. (1972) Decreased blood PCO_2 in the ovine foetus during hypothermia: implications for increased placental blood flow. *Experientia* 28, 801-802.
266. Hall, K. and Olin, P. (1972) Sulphation factor activity and growth rate during long-term treatment of patients with pituitary dwarfism with human growth hormone. *Acta Endocrinol.* 69, 417-433.
267. Hamburgh, M. (1968) An analysis of the action of thyroid hormone on development based on in vivo and in vitro studies. *J. Gen. Comp. Endocrinol.* 10, 198-213.
268. Hamburgh, M., Lynn, E. and Weiss, E.P. (1964) Analysis of the influence of thyroid hormone on prenatal and postnatal maturation of the rat. *Anat. Rec.* 150, 147-162.
269. Hamerton, J.L. (1968) Significance of sex chromosome derived heterochromatin in mammals. *Nature* 219, 910-914.
270. Harris, G.W. (1964) Sex hormones, brain development and brain function. *Endocrinology* 75, 627-648.
271. Harrison, F.A., Heap, R.B. and Linzell, J.L. (1968) Ovarian function in the sheep after autotransplantation of the ovary and uterus to the neck. *J. Endocrinol.* 40, xiii.
272. Hedge, G.A., Yates, M.B., Marcus, R. and Yates, F.E. (1966) Site of action of vasopressin in causing corticotrophin release. *Endocrinology* 79, 328-340.
273. Heggestad, C.B. and Wells, L.J. (1954) Lack of compensatory changes in the developing thyroid in foetal rats from hypophysectomised mothers. *Anat. Rec.* 118, 389-390.
274. Heggestad, C.B. and Wells, L.J. (1965) Experiments on the contribution of somatotrophin to prenatal growth in the rat. *Acta Anat.* 60, 348-361.
275. Heins, J.N., Garland, J.T. and Daughaday, W.H. (1970) Incorporation of ^{35}S-sulfate into rat cartilage explants in vitro: effects of aging on responsiveness to stimulation by sulfation factor. *Endocrinology* 87, 688-692.
276. Hendricks, C.H. (1954) The neurohypophysis in pregnancy. *Obstet. Gynecol. Surv.* 9, 323.

277. Hennen, G., Pierce, J.G. and Freychet, P. (1969) Human chorionic thyrotrophin: further characteristics and study of its secretion during pregnancy. *J. Clin. Endocrinol.* 29, 581-594.
278. Herbst, A.L., Ulfelder, H. and Poskanzer, D.C. (1971) Adenocarcinoma of the vagina. *New Engl. J. Med.* 284, 878-881.
279. Hershman, M. and Starnes, W.R. (1968) Extraction of a thyrotrophin from the human placenta. In: *Excerpta Medica Int. Cong. Ser.* No. 157, Abstract 496. Third International Congress of Endocrinology, Mexico, p. 199.
280. Heymann, M.A. and Rudolph, A.M. (1967) Effect of exteriorization of the sheep fetus on its cardiovascular function. *Circulat. Res.* 21, 741-745.
281. Hill, E.C. (1973) Clear cell carcinoma of the cervix and vagina in young women. *Am. J. Obstet. Gynecol.* 116, 470-481.
282. Himsworth, R.L., Carmel, R.W. and Frantz, A.G. (1972) The location of the chemotransmitter controlling growth hormone secretion during hypoglycaemia in primates. *Endocrinology* 91, 217-226.
283. Hindson, J.C., Schofield, B.M. and Turner, C.B. (1967) The effect of a single dose of stilboestrol on cervical dilatation in pregnant sheep. *Res. Vet. Sci.* 8, 353-360.
284. Hindson, J.C., Schofield, B.M. and Turner, C.B. (1968) Parturient pressures in the ovine uterus. *J. Physiol.* 195, 19-28.
285. Hindson, J.C., Schofield, B.M. and Turner, C.B. (1968) Some factors affecting dilatation of the ovine cervix. *Res. Vet. Sci.* 9, 474-480.
286. Hindson, J.C., Schofield, B.M., Turner, C.B. and Wolff, H.S. (1965) Parturition in the sheep. *J. Physiol.* 181, 560-567.
287. Hindson, J.C. and Ward, W.R. (1973) Myometrial studies in the pregnant sheep. In: *Pregnancy and Parturition.* Ed.: Pierrepoint, C.G. (Tenovus Library Series, Alpha Omega Alpha, Cardiff) pp. 153-162.
288. Hiroshige, T. and Sato, T. (1971) Changes in hypothalamic content of corticotrophin-releasing activity following stress during neonatal maturation in the rat. *Neuroendocrinology* 7, 257-270.
289. Hirvonen, L. and Lybeck, H. (1956) On the permeability of guinea pig placenta for thyroxine. *Acta Physiol. Scand.* 36, 18-22.
290. Hodari, A.A. (1968) The contribution of the fetal kidney to experimental hypertensive disease of pregnancy. *Am. J. Obstet. Gynecol.* 101, 17-22.
291. Hodges, R.E., Evans, T.C., Bradbury, J.T. and Kestell, W.C. (1955) The accumulation of radioactive iodine by human foetal thyroids. *J. Clin. Endocrinol. Metab.* 15, 661-667.
292. Hoet, J.J. (1974) In: Size at Birth. *CIBA Found. Symp.* 27, Eds.: Elliott, K. and Knight, J. (Elsevier, Amsterdam) p. 201.
293. Hofmann, K. (1974) Relations between chemical structure and function of adrenocorticotrophin and melanocyte-stimulating hormones. In: *Handbook of Physiology*, Section 7, Endocrinology, Vol. IV. The pituitary gland and its neuroendocrine control Part 2. Eds.: Knobil, E. and Sawyer, W.H. (Am. Physiol. Soc., Washington, D.C.) pp. 29-58.
294. Hofmann, K., Wingender, W. and Finn, F.M. (1970) Correlation of adrenocorticotropic activity of ACTH analogs with degree of binding to an adrenal cortical particulate preparation. *Proc. Nat. Acad. Sci.* 67 (2) 829-836.
295. Holley, D.C., Beckman, D.A. and Evans, J.W. (1975) Effect of confinement on the circadian rhythm of ovine cortisol. *J. Endocrinol.* 65, 147-148.

296. Holley, D.C. and Evans, J.W. (1974) Effect of confinement on ovine glucose and immunoreactive insulin circadian rhythms. *Am. J. Physiol.* 226, 1457-1461.
297. Holm, L.W., Parker, H.R. and Galligan, S.J. (1961) Adrenal insufficiency in postmature calves. *Am. J. Obstet. Gynecol.* 81, 1000-1008.
298. Holt, A.B., Kerr, G.R. and Cheek, D.B. (1975) Prenatal hypothyroidism and brain composition. In: *Fetal and Postnatal Cellular Growth. Hormones and Nutrition.* Ed.: Cheek, D.B. (Wiley) pp. 141-154.
299. Holt, P.G. and Oliver, I.T. (1968) Factors affecting the premature induction of tyrosine amino transferase in foetal rat liver. *Biochem. J.* 108, 333-338.
300. Honnebier, W.J., Jobsis, A.C. and Swaab, D.F. (1974) The effect of hypophysial hormones and human chorionic gonadotrophin (HCG) on the anencephalic fetal adrenal cortex and on parturition in the human. *J. Obstet. Gynaecol. Br. Commonw.* 81, 423-438.
301. Honnebier, W.J. and Swaab, D.F. (1973) The influence of anencephaly upon intrauterine growth of fetus and placenta and upon gestation length. *J. Obstet. Gynaecol. Br. Commonw.* 80, 577-588.
302. Honnebier, W.J. and Swaab, D.F. (1974) Influence of α-melanocyte stimulating hormone (MSH), growth hormone (GH) and fetal brain extracts on intrauterine growth of fetus and placenta in the rat. *J. Obstet. Gynaecol. Br. Commonw.* 81, 439-447.
303. Hope, D.B. and Pickup, J.C. (1974) Neurophysins. In: *Handbook of Physiology.* Section 7. Endocrinology. Volume IV. Part I (Am. Physiol. Soc.) pp. 173-189.
304. Hopkins, P.S. and Thorburn, G.D. (1971) Plasma thyroxine and cortisol concentrations of the foetal lamb. *Fourth Asia and Oceania Congress in Endocrinology* (University of Auckland) Abstract 170.
305. Hopkins, P.S. and Thorburn, G.D. (1972) The effects of foetal thyroidectomy on the development of the ovine foetus. *J. Endocrinol.* 54, 55-66.
306. Hopkins, P.S., Wallace, A.L.C. and Thorburn, G.D. (1975) Thyrotrophin concentrations in the plasma of cattle, sheep and foetal lambs as measured by radioimmunoassay. *J. Endocrinol.* 64, 371-387.
307. Hoppenstein, J.M., Miltenberger, F.W. and Moran, W.H. (1968) The increase in blood levels of vasopressin in infants during birth and surgical procedures. *Surg. Gynecol. Obstet.* 127, 966.
308. Horrobin, D.F. (1974) *Prolactin 1974.* (Medical and Technical Publishing Co. Ltd.) Chapter 16.
309. Horrobin, D.F., Manku, M.S., Nassar, B. and Evered, D. (1973) Prolactin and fluid and electrolyte balance. In: *Human prolactin. Proc. Int. Symp. Human Prolactin.* Eds.: Pasteels, J.L. and Robyn, C. (Elsevier, Amsterdam) pp. 152-155.
310. Hsueh, A.J.W., Peck, E.J. and Clark, J.H. (1975) Progesterone antagonism of the oestrogen receptor and oestrogen-induced uterine growth. *Nature* 254, 339-341.
311. Hüfner, M., Hesch, R.D., Luders, D. and Heinrich, U. (1973) Plasma T3 at the end of pregnancy in cord blood and during the first days of life. *Acta Endocrinol. Kbh.* suppl. 173, 16.
312. Hull, D. (1975) Storage and supply of fatty acids before and after birth. *Br. Med. Bull.* 31 (1) 32-36.
313. Hwang, U.K. and Wells, L.J. (1959) Hypophysis–thyroid system in the foetal rat: thyroid after hypophyseoprivia, thyroxine, triiodothyronine, thyrotropin and growth hormone. *Anat. Rec.* 134, 125-141.

314. Hyman, A.I. and Towell, M.F. (1968) Effects of catecholamine depletion on metabolic response to cold in the newborn guinea pig. *Am. J. Physiol.* 214, 691-694.
315. Hyppa, M. (1972) Hypothalamic monoamines in human fetuses. *Neuroendocrinology* 9, 257-266.
316. Ibbertson, H.K. (1974) Goitre and cretinism in the high Himalayas. *New Zealand Med. J.* 80, 484-488.
317. Ibbertson, H.K., Gluckman, P.D., Croxson, M.S. and Strang, L.J.W. (1974) Goiter and cretinism in the Himalayas: a reassessment. In: *Endemic Goiter and Cretinism: continuing threats to World Health*. PAHO Sci. Publ. No. 292. Ed.: Dunn, J.T. and Medeiros-Neto, G.A. (WHO Washington D.C.).
318. Igic, R., Erdos, E.G., Yeh, H.S.J., Sorrells, K. and Nakajima, T. (1972) Angiotensin 1 converting enzyme of the lung. *Circulat. Res.* 31, suppl. 2, 11-51 - 11-61.
319. Jack, P.M.B., Albano, J.D.M., Brown, B.L., Gould, R.P., Joseph, T. and Nathanielsz, P.W. (1975) Do hormones induce their own receptor in foetal life? The suppression of ACTH-sensitive adenylate cyclase in the foetal rabbit by the administration of cortisol. *J. Endocrinol.* 64, 67-68P.
320. Jack, P.M.B. and Nathanielsz, P.W. (1974) Inhibition of the oxytocic action of prostaglandin $F_{2\alpha}$ on the pregnant rabbit uterus by intra-aortic infusion of theophylline in vivo. *J. Endocrinol.* 62, 171-172.
321. Jack, P.M.B. and Nathanielsz, P.W. (1974) The effect of theophylline on oxytocin-induced contractions in the chronically catheterized pregnant rabbit. *Experientia* 30, 1218-1219.
322. Jack, P.M.B., Nathanielsz, P.W., Rees, L.H. and Thomas, A.L. (1975) Plasma adrenocorticotrophin concentrations during induction of parturition by the intravascular infusion of physiological amounts of cortisol into the sheep foetus. *J. Physiol.* 245, 76-78P.
323. Jack, P.M.B., Nathanielsz, P.W., Thomas, A.L. and Steven, D.H. (1975) Ultrastructural changes in the placenta of the ewe following foetal infusion of cortisol. *Quart. J. Exp. Physiol.* 60, 171-179.
324. Jackson, B.T. and Piasecki, G.J. (1969) Foetal secretion of glucocorticoids. *Endocrinology* 85, 875-880.
325. Jacobson, A.G. and Brent, R.L. (1959) Radioiodine concentration by the foetal mouse thyroid. *Endocrinology* 65, 408-416.
326. James, E., Meschia, G. and Battaglia, F.C. (1971) Arterio-venous differences of free acids and glycerol in the ovine umbilical circulation. *Proc. Soc. Exp. Biol. Med.* 138, 823-826.
327. Jochle, W. (1973) Corticosteroid-induced parturition in domestic animals. *Annu. Rev. Pharmacol.* 13, 33-55.
328. Johannisson, E. (1968) Foetal adrenal cortex in human: its ultrastructure at different stages of development and in different functional states. *Acta Endocrinol.* 58, suppl. 130, 7-107.
329. John, B.M. and Pierrepoint, C.G. (1975) Demonstration of an active C17-20 lyase in the normal sheep placenta. *J. Reprod. Fertil.* 43, 559-562.
330. Johnson, P., Jones, C.T., Kendall, J.Z., Ritchie, J.W.K. and Thorburn, G.D. (1975) ACTH and the induction of parturition in sheep. *J. Physiol.* 252, 64-66P.
331. Jones, C.T., Luther, E., Ritchie, J.W.K. and Worthington, D. (1975) The clearance of ACTH from the plasma of adult and fetal sheep. *Endocrinology* 96, 231-234.

332. Jones, C.T. and Robinson, R.O. (1975) Plasma catecholamines in foetal and adult sheep. *J. Physiol.* 248, 15-33.
333. Josimovich, J.B., Ladman, A.J. and Deane, H.W. (1954) A histophysiological study of the developing adrenal cortex of the rat during fetal and early postnatal stages. *Endocrinology* 54, 627-639.
334. Josso, N. (1973) *Abstracts of IV International Congress of Endocrinology*. Washington D.C., 1972. Excerpta Medica International Congress Series 256, 153.
335. Jost, A. (1947) Expériences de décapitation de l'embryon de lapin. *CR Acad. Sci. Paris* 225, 322-324.
336. Jost, A. (1959) Action du propylthiouracile sur la thyroide du foetus de lapin intact, decapité ou injecté de thyroxine. *CR Hebd. Séances Mém. Soc. Biol.* 153, 1900-1902.
337. Jost, A. (1966) Problems of fetal endocrinology: the adrenal glands. *Rec. Prog. Horm. Res.* 22, 541-574.
338. Jost, A. (1973) Does the foetal hypophyseal-adrenal system participate in delivery in rats and rabbits? In: *Foetal and Neonatal Physiology*. Proc. Sir Joseph Barcroft Centenary Symp. Eds.: Comline, R.S., Cross, K.W., Dawes, G.D. and Nathanielsz, P.W. (Cambridge University Press) pp. 589-593.
339. Jost, A. (1976) Sexual differentiation. In: *Fetal Physiology and Medecine*. Eds.: Beard, R.W. and Nathanielsz, P.W. (W.B. Saunders).
340. Jost, A. and Cohen, A. (1966) Significance of the atrophy of the fetal adrenal glands of the rat induced by hypophysectomy (decapitation). *Dev. Biol.* 14, 154-168.
341. Jost, A., Dupouy, J.P. and Geloso-Meyer, A. (1970) The hypothalamo-hypophyseal relationships in the fetus. In: *The Hypothalamus*. Proc. Workshop Conf. Integration of Endocrine and Non-endocrine Mechanisms in the Hypothalamus, Stresa 1969. Eds.: Martini, L., Motta, M. and Fraschini, E. (Academic Press, New York) pp. 605-615.
342. Jost, A., Dupouy, J.P. and Rieutort, M. (1974) The ontogenetic development of hypothalamo-hypophyseal relations. In: *Integrative Hypothalamic Activity*. Eds.: Swaab, D.F. and Schadé, J.P. *Progress in Brain Research*, 41. (Elsevier, Amsterdam) pp. 209-219.
343. Jost, A. and Picon, L. (1970) Hormonal control of fetal development and metabolism. *Adv. Metab. Disorders* 4, 123-184.
344. Jost, A., Vigier, B. and Prepin, J. (1972) Free martins in cattle: the first steps in sexual organogenesis. *J. Reprod. Fertil.* 29, 349-379.
345. Kaiser, I.H. and Halberg, F. (1962) Circadian periodic aspects of birth. *Ann. NY Acad. Sci.* 98, 1056-1058.
346. Kaiser, W. and Bygrave, F.G. (1969) Stimulation of phospholipid synthesis in isolated rat liver mitochondria after treatment in vivo with triiodothyronine. *Eur. J. Biochem.* 11, 93-96.
347. Kajihara, A., Kojima, A., Onaya, T., Takemura, Y. and Yamada, T. (1972) Placental transport of thyrotrophin releasing factor in the rat. *Endocrinology* 90, 592-594.
348. Kao, C.Y. (1973) Ionic currents in a pregnant myometrium. In: *Foetal and Neonatal Physiology*. Proc. Sir Joseph Barcroft Centenary Symp. Eds.: Comline, R.S., Cross, K.W., Dawes, G.S. and Nathanielsz, P.W. (Cambridge University Press) pp. 584-588.

349. Kaplan, S.L., Grumbach, M.M. and Shepard, T.H. (1972) The ontogenesis of human fetal hormones. I. Growth Hormone and Insulin. *J. Clin. Invest.* 51, 3080-3093.
350. Kaplan, S.L., Gurpide, E., Sciarra, J.J. and Grumbach, M.M. (1968) Metabolic clearance rate and production rate of chorionic growth hormone – prolactin in late pregnancy. *J. Clin. Endocrinol. Metab.* 28, 1450-1460.
351. Karim, S.M.M. (1972) Prostaglandin and reproduction: physiological roles and clinical uses of prostaglandins in relation to human reproduction. In: *Prostaglandins: Progress in Research.* Ed.: Karim, S.M.M. (Medical and Technical Press, Oxford).
352. Karim, S.M.M. and Devlin, J. (1967) Prostaglandin content of amniotic fluid during pregnancy and labour. *J. Obstet. Gynaecol. Br. Commonw.* 74, 230-234.
353. Karim, S.M.M., Hillier, K. and Devlin, J. (1968) Distribution of prostaglandins E_1, E_2, $F_{1\alpha}$ and $F_{2\alpha}$ in some animal tissues. *J. Pharm. Pharmacol.* 20, 749-753.
354. Karsch, F.J., Krey, L.C., Weick, R.F., Dierschke, D.J. and Knobil, E. (1973) Functional luteolysis in the rhesus monkey – the role of estrogen. *Endocrinology* 92, 1148-1152.
354a Keirse, M.J.N.C., Flint, A.P.F. and Turnbull, A.C. (1974) F prostaglandins in amniotic fluid during pregnancy and labour. *J. Obstet. Gynaecol. Br. Commonw.* 81, 131-135.
354b Keirse, M.J.N.C., Hicks, B.R. and Turnbull, A.C. (1976) Metabolism of prostaglandin $F_{2\alpha}$ in fetal and maternal cotyledons of sheep. *J. Reprod. Fertil.* 46, 417-420.
354c Keirse, M.J.N.C., Hicks, B.R. and Turnbull, A.C. (1975) Comparison of intrauterine metabolism of prostaglandin $F_{2\alpha}$ in ovine and human pregnancy. *J. Endocrinol.* 67, 24-25P.
354d Keirse, M.J.N.C. and Turnbull, A.C. (1976) The fetal membranes as a possible source of amniotic fluid prostaglandins. *Br. J. Obstet. Gynaecol.* 83, 146-151.
355. Kelly, P.A., Robertson, H.A. and Friesen, H.G. (1974) Temporal pattern of placental lactogen and progesterone secretion in sheep. *Nature* 248, 435-437.
356. Kelly, P.A., Shiu, R.P.C., Robertson, M.C. and Friesen, H.G. (1975) Characterization of rat chorionic mammotrophin. *Endocrinology* 96, 1187-1195.
357. Kendall, J.W. and Roth, J.G. (1969) Adrenocortical function in monkeys after forebrain removal or pituitary stalk section. *Endocrinology* 84, 686-691.
358. Kennedy, P.C., Kendrick, J.W. and Stormont, C. (1957) Adenohypophyseal aplasia, an inherited defect associated with abnormal gestation in Guernsey cattle. Cornell Vet. 47, 160-178.
359. Kerr, G.R., Tyson, I.B., Allen, J.R., Wallace, J.H. and Scheffler, G. (1972) Deficiency of thyroid hormone and development of the fetal rhesus monkey. I. Effect on physical growth, skeletal maturation and biochemical measures of thyroid function. *Biol. Neonate* 21, 282-295.
360. Kipnis, D.M., Hertelendy, F. and Machlin, L.J. (1969) Studies of growth hormone secretion. In: *Progress in Endocrinology.* Ed.: Gual, C. Excerpta Medica Foundation, Amsterdam, 1969, pp. 601-609.
361. Kirsch, R.E. (1938) A study on the control of length of gestation of the rat with notes on maintenance and termination of pregnancy. *Am. J. Physiol.* 122, 86-93.
362. Kirton, K.T., Pharriss, B.B. and Forbes, A.D. (1970) Luteolytic effects of prostaglandin $F_{2\alpha}$ in primates. *Proc. Soc. Exp. Biol. Med.* 133, 314-319.

363. Kittinger, G.W. (1973) The regulation of cortisol levels in fetal plasma. *Primate News* 11, 2-5.
364. Kittinger, G.W. (1974) Feto-maternal production and transfer of cortisol in the rhesus (*Macaca mulatta*). *Steroids* 23, 229-243.
365. Kittinger, G.W., Beamer, N.B., Hagemenas, F., Hill, J.D., Baughman, W.L. and Ochsner, A.J. (1972) Evidence for autonomous pituitary-adrenal function in the near-term fetal rhesus (*Macaca mulatta*). *Endocrinology* 91, 1037-1044.
366. Klaiber, E.L., Lloyd, C., Solomon, A. and Broverman, D.M. (1968) Augmentation of the effects of testosterone proprionate by cortisol in immature male rats. *Endocrinology* 83, 387-389.
367. Klevit, H.D. (1966) Foetal-placental-maternal interrelations involving steroid hormones. *Paediatr. Clin. North Am.* 13 (1) 59-71.
368. Knobil, E. and Josimovich, J.B. (1958) Placental transfer of thyrotropic hormone, thyroxine, triiodothyronine and insulin in the rat. *Ann. NY Acad. Sci.* 75, 895-904.
369. Koch, H.C., Reighert, W., Stolte, L., Van Kesses, H. and Seelen, J. (1966) Placental thyroxine transfer and fetal thyroxine utilization. *Acta Physiol. Pharmacol. Neerl.* 13, 363-365.
370. Koutras, D.A., Berman, M., Sfontouris, J., Rigopoulos, G.A., Koukoulommati, A.S. and Malames, B. (1970) Endemic goitre in Greece: thyroid hormone kinetics. *J. Clin. Endocrinol. Metab.* 24, 857-862.
371. Krieger, D.T. (1974) Effect of neonatal hydrocortisone on corticosteroid circadian periodicity, responsiveness to ACTH and stress in prepuberal and adult rats. *Neuroendocrinology* 16, 355-363.
372. Kuehl, F.A. (1974) Prostaglandins, cyclic nucleotides and cell function. *Prostaglandins* 5, 325-338.
373. Kuehl, F.A., Cirillo, V.J., Ham, E.A. and Humes, J.L. (1973) Regulatory role of the prostaglandins on the cyclic 3',5'-AMP system. In: *Advances in the Biosciences*, 9, Int. Conf. Prostaglandins. Ed.: Raspé, G. (Pergamon Press, Vieweg) pp. 155-172.
374. Kumaresan, P. and Turner, C.W. (1966) Effect of neonatal administration of testosterone propionate upon thyroid hormone secretion rate in female rats. *Endocrinology* 79, 1009-1010.
375. Lacomme, M., Tuchmann-Duplessis, H. and Mercier-Parot, L. (1964) Neuroendocrine control of the adrenal cortex of the human fetus. *Gynécol. Obstét.* (Paris) 63, 421-440.
376. Lamming, G.E., Moseley, S.R. and McNeilly, J.R. (1974) Prolactin release in the sheep. *J. Reprod. Fertil.* 40, 151-168.
377. Lanman, J.T. (1968) Delays during reproduction and their effects on the embryo and fetus. *New Engl. J. Med.* 278, 993-999, 1047-1054, 1092-1099.
378. Lanman, J.T., Thau, R., Sundaram, K., Brinson, A. and Bonk, R. (1975) Ovarian and placental origins of plasma progesterone following fetectomy in monkeys (*Macaca mulatta*). *Endocrinology* 96, 591-597.
379. Lardy, H.A. and Feldoth, G. (1951) Metabolic effects of thyroxine in vitro. *Ann. NY Acad. Sci.* 54, 636-647.
380. Laron, Z., Pertzelan, A., Karp, M., Kowaldo-Silbergeld, A. and Daughaday, W.H. (1971) Administration of growth hormone to patients with familial dwarfism with high immunoreactive growth hormone: measurement of sulfation factor, metabolic and linear growth responses. *J. Clin. Endocrinol. Metab.* 33, 332-342.

381. Laron, Z., Pertzelan, A., Mannheimer, S., Goldman, J. and Guttmann, S. (1966) Lack of placental transfer of human growth hormone. *Acta Endocrinol.* 53, 687-692.
382. Lascelles, A.K. (1959) The time of appearance of ossification centres in the Peppin-type Merino. *Aust. J. Zool.* 7, 79-86.
383. Lascelles, A.K. and Setchell, B.P. (1959) Hypothyroidism in sheep. *Aust. J. Biol. Sci.* 12, 445-465.
384. Legros, J.J. and Franchimont, P. (1972) Human neurophysine blood levels under normal, experimental and pathological conditions. *Clin. Endocrinol.* 1, 99-113.
385. Leung, K. and Munck, A. (1975) Peripheral actions of glucocorticoids. *Annu. Rev. Physiol.* 37, 245-272.
386. Liggins, G.C. (1968) Premature parturition after infusion of corticotrophin or cortisol into foetal lambs. *J. Endocrinol.* 42, 323-329.
387. Liggins, G.C. (1969) The foetal role in the initiation of parturition in the ewe. In: *Foetal Autonomy*. Ciba Foundation Symposium. Ed.: Wolstenholme, G.E.W. and O'Connor, M. pp. 218-244.
387a Liggins, G.C. (1969) Premature delivery of foetal lambs infused with glucocorticoids. *J. Endocrinol.* 45, 515-523.
388. Liggins, G.C. (1974) Parturition in the sheep and the human. In: *Physiology and Genetics of Reproduction*. Part B. Eds.: Coutinho and Fuchs. (Plenum Press) pp. 423-443.
389. Liggins, G.C. (1974) The influence of the fetal hypothalamus and pituitary on growth. In: *Size at Birth*. Ciba Foundation Symposium 27. Eds.: Elliott, K. and Knight, J. pp. 165-183.
390. Liggins, G.C. (1976) The Drive to Growth. In: *Fetal Physiology and Medicine*. Eds.: Beard, R.W. and Nathanielsz, P.W. (W.B. Saunders).
391. Liggins, G.C., Fairclough, R.J., Grieves, S.A., Kendall, J.Z. and Knox, B.S. (1973) The mechanism of initiation of parturition in the ewe. *Rec. Prog. Horm. Res.* 29, 111-159.
391a Liggins, G.C., Grieves, S.A., Kendall, J.Z. and Knox, B.S. (1972) The physiological roles of progesterone, oestradiol-17β and prostaglandin $F_{2\alpha}$ in the control of ovine parturition. *J. Reprod. Fertil.* suppl. 16, 85-103.
392. Liggins, G.C., Holm, L.W. and Kennedy, P.C. (1966) Prolonged pregnancy following surgical lesions of the foetal lamb pituitary. *J. Reprod. Fertil.* 12, 419.
393. Liggins, G.C. and Howie, R.N. (1974) The prevention of RDS by maternal steroid therapy. In: *Modern Perinatal Medicine*. Ed.: Gluck, L. (Year Book Medical Publishers, Chicago).
394. Liggins, G.C. and Howie, R. (1976) In: Proceeding of the 70th Ross Conference (In press).
395. Liggins, G.C. and Kennedy, P.C. (1968) Effects of electrocoagulation of the foetal lamb hypophysis on growth and development. *J. Endocrinol.* 40, 371-381.
396. Liggins, G.C., Kennedy, P.C. and Holm, L.W. (1967) Failure of initiation of parturition after electrocoagulation of the pituitary of the foetal lamb. *Am. J. Obstet. Gynecol.* 98, 1080-1086.
397. Lincoln, D.W. (1973) Milk ejection during alcohol anaesthesia in the rat. *Nature* 243, 227-229.
398. London, W.T., Money, W.L. and Rawson, R.W. (1963) Placental transport of I^{131}-labelled thyroxine and triiodothyronine in the guinea pig. *Endocrinology* 73, 205-209.

399. MacDonald, P.C., Schultz, F.M., Duenhoelter, J.H., Gant, N.F., Jimenez, J.M., Pritchard, J.A., Porter, J.C. and Johnston, J.M. (1974) Initiation of human parturition. I. Mechanism of action of arachidonic acid. *J. Obstet. Gynaecol.* 44, 629-636.
400. Machlin, L.J., Takahashi, T., Horino, M., Hertelendy, F., Gordon, R.S. and Kipnis, D. (1968) Regulation of growth hormone secretion in non-primate species. In: *Proceedings of the 1st International Symposium on Growth Hormone.* Milan. (Excerpta Medica Foundation International Congress Series 158, Amsterdam) pp. 292-305.
401. McCann, S.M. and Brobeck, J.E. (1954) Evidence for a role of the supraoptico-hypophyseal system in regulation of adrenocorticotrophin secretion. *Proc. Soc. Exp. Biol. Med.* 87, 318.
402. McNatty, K.P. and Cashmore, M. (1972) Studies on diurnal variation of plasma cortisol levels in sheep. *New Zealand Med. J.* 76, 115.
403. Madill, D. and Bassett, J.M. (1973) Corticosteroid release by adrenal tissue from foetal and newborn lambs in response to corticotrophin stimulation in a perifusion system in vitro. *J. Endocrinol.* 58, 75-87.
404. Malinowska, K.W., Chan, W.S., Nathanielsz, P.W. and Hardy, R.N. (1974) Plasma adrenocorticosteroid changes during thyroxine-induced accelerated maturation of the neonatal rat intestine. *Experientia* 30, 61.
405. Malinowska, K.W., Hardy, R.N. and Nathanielsz, P.W. (1972) Plasma adrenocorticosteroid concentrations immediately after birth in the rat, rabbit and guinea pig. *Experientia* 28, 1366-1367.
405a Malinowska, K.W. and Nathanielsz, P.W. (1974) Plasma aldosterone, cortisol and corticosterone in the newborn guinea pig. *J. Physiol.* 236, 83-93.
406. Malpas, P. (1933) Postmaturity and malformations of the foetus. *J. Obstet. Gynaecol. Br. Empire* 40, 1046-1053.
407. Martal, J. and Dijane, J. (1975) Purification of a lactogenic hormone in sheep placenta. *Biochem. Biophys. Res. Commun.* 65, 770-778.
408. Martin, M.J., Chard, T. and Landon, J. (1972) Development of radioimmunoassay for bovine neurophysin. *J. Endocrinol.* 52, 481-495.
409. Mati, J.K.G., Horrobin, D.F. and Bramley, P.S. (1973) Induction of labour in sheep and in humans by single doses of corticosteroids. *Br. Med. J.* 2, 149-151.
410. Matsumoto, K., Kotoh, K., Miyata, J. and Kurachi, K. (1965) Effect of 2-methyl-1, 2-bis-(3-pyridyl)-1-propanone (Su 4885) on oestrogen biosynthesis in vivo. *J. Endocrinol.* 33, 317-318.
411. Matthies, D.L. (1967) Studies of the luteotropic and mammotropic factor found in trophoblast and maternal peripheral blood of the rat at mid-pregnancy. *Anat. Rec.* 159, 55-67.
412. Mellor, D.J. and Matheson, I.C. (1975) Chronic catheterisation of the aorta and umbilical vessels of fetal sheep. *Res. Vet. Sci.* 18, 221-223.
412a Mellor, D.J., Matheson, I.C., Small, J. and Wright, H. (1976) Plasma thyroxine concentrations in ewes and their fetuses during the last six weeks of pregnancy. *Res. Vet. Sci.* 21, 102-103.
413. Mellor, D.J. and Pearson, R.A. (1974) Changes in ionic composition of allantoic fluid during adrenocorticotrophin infusion into fetal sheep. *Res. Vet. Sci.* 16, 108-109.
414. Mellor, D.J. and Slater, J.S. (1971) Daily changes in amniotic and allantoic fluid

during the last three months of pregnancy in conscious, unstressed ewes, with catheters in their foetal fluid sacs. *J. Physiol.* 217, 573-604.

415. Mellor, D.J. and Slater, J.S. (1972) Daily changes in foetal urine and relationships with amniotic and allantoic fluid and maternal plasma during the last two months of pregnancy in conscious, unstressed ewes with chronically implanted catheters. *J. Physiol.* 227, 503-525.

416. Mellor, D.J. and Slater, J.S. (1973) The composition of maternal plasma and fetal urine after feeding and drinking in chronically catheterised ewes during the last two months of pregnancy. *J. Physiol.* 234, 519-531.

417. Mellor, D.J., Slater, J.S. and Cockburn, F. (1971) Effects of antibiotic treatment on the composition of sheep foetal fluids. *Res. Vet. Sci.* 12, 521-526.

418. Mellor, D.J., Slater, J.S. and Matheson, I.C. (1975) Effect of changes in ambient temperature on maternal plasma and allantoic fluid from chronically catheterised ewes during the last two months of pregnancy. *Res. Vet. Sci.* 18, 219-221.

419. Meschia, G., Cotter, J.R., Breathnach, C.S. and Barron, D.H. (1965) The haemoglobin, oxygen, carbon dioxide and hydrogen ion concentrations in the umbilical bloods of sheep and goats as sampled via indwelling plastic catheters. *Quart. J. Exp. Physiol.* 50, 185-195.

420. Migeon, C.J., Prystowsky, H., Grumbach, M.M. and Byron, M.C. (1956) Placental passage of 17-hydroxycorticosteroids: comparison of the levels in maternal and foetal plasma and effects of ACTH and hydrocortisone administration. *J. Clin. Invest.* 35, 488-493.

421. Milic, A.B. and Adamsons, K. (1969) The relationship between anencephaly and prolonged pregnancy. *J. Obstet. Gynaecol. Br. Commonw.* 76, 102-111.

422. Mintz, D.H., Chez, R.A. and Horger, E.O. (1969) Fetal insulin and growth hormone metabolism in the sub-human primate. *J. Clin. Invest.* 48, 176-186.

423. Miyabo, S. and Hisado, T. (1975) Sex difference in ontogenesis of circadian adrenocortical rhythm in cortisone-primed rats. *Nature* 256, 590-592.

424. Moger, W.H. and Geschwind, J.J. (1971) Plasma prolactin levels in fetal sheep. *Experientia* 27, 1479-1480.

425. Mongkonpunya, K., Lin, Y.C., Noden, P.A., Oxender, W.D. and Hafs, H.D. (1975) Androgens in the bovine fetus and dam. *Proc. Soc. Exp. Biol. Med.* 148, 489-493.

426. Mossman, R.G. and Conrad, J.T. (1969) Oxytocin and modulating effects of water-soluble hydrocortisone and methyl prednisolone upon in vitro contractions of myometrium. *Am. J. Obstet. Gynecol.* 105, 897-908.

427. Mott, J.C. (1975) The place of the renin–angiotensin system before and after birth. *Br. Med. Bull.* 31 (1) 44-49.

428. Motta, M., Mangili, G. and Martini, L. (1965) A 'short' feedback loop in the control of ACTH secretion. *Endocrinology* 77, 392-395.

428a Mueller-Heubach, E., Myers, R.E. and Adamsons, K. (1972) Effects of adrenalectomy on pregnancy length in the rhesus monkey. *Am. J. Obstet. Gynecol.* 112, 221-226.

429. Mulay, S., Giannopoulos, G. and Solomon, S. (1973) Corticosteroid levels in the mother and fetus of the rabbit during gestation. *Endocrinology* 93, 1342-1348.

430. Murphy, B.E.P. (1973) Steroid arteriovenous differences in umbilical cord plasma: evidence of cortisol production by the human fetus in early gestation. *J. Clin. Endocrinol. Metab.* 36, 1037-1038.

431. Murphy, B.E.P. (1973) Does the human fetal adrenal play a role in parturition? *Am. J. Obstet. Gynecol.* 115, 521-525.
432. Murphy, B.E.P., Clark, S.J., Donald, I.R., Pinsky, M. and Vedady, D. (1974) Conversion of maternal cortisol to cortisone during placental transfer to the human fetus. *Am. J. Obstet. Gynecol.* 118, 538-541.
433. Murphy, B.E.P. and Diez D'Aux, R.C. (1972) Steroid levels in the human fetus: cortisol and cortisone. *J. Clin. Endocrinol.* 35, 678-683.
434. Murphy, B.E.P., Patrick, J. and Denton, R.L. (1975) Cortisol in amniotic fluid during human gestation. *J. Clin. Endocrinol. Metab.* 40, 164-167.
435. Myant, N.B. (1958) Passage of thyroxine and triiodothyronine from mother to fetus in pregnant women. *Clin. Sci.* 17, 75-79.
436. Myers, R.E. (1975) Maternal psychological stress and fetal asphyxia: A study in the monkey. *Am. J. Obstet. Gynecol.* 122, 47-59.
437. Nancarrow, C.D. and Seamark, R.F. (1968) Progesterone metabolism in fetal blood. *Steroids* 12, 367-380.
438. Nataf, B.M. and Chaikoff, I.L. (1964) Effect of injected thyrotropic hormone on the in vitro uptake and metabolism of ^{131}I by thyroid glands of foetal and newborn rats. *Endocrinology* 75, 547-553.
438a Nathanielsz, P.W. (1968) The effect of acute starvation for 48 hours on deiodination of a tracer dose of ^{131}I-thyroxine in the rat compared with the effect of hydrocortisone. *J. Physiol.* 194, 80-81P.
439. Nathanielsz, P.W. (1975) Thyroid function in the fetus and newborn mammal. *Br. Med. Bull.* 31 (1) 51-56.
440. Nathanielsz, P.W. (1977) The endocrinology of parturition. *Annu. Rev. Physiol.* (In press).
441. Nathanielsz, P.W., Abel, M. and Smith, G.W. (1973) Hormonal factors in parturition in the rabbit. In: *Foetal and Neonatal Physiology*. Proc. Sir Joseph Barcroft Centenary Symp. Eds.: Comline, R.S., Cross, K.W., Dawes, G.S. and Nathanielsz, P.W. (Cambridge University Press).
442. Nathanielsz, P.W., Comline, R.S., Silver, M. and Paisey, R.B. (1972) Cortisol metabolism in the fetal and neonatal sheep. *J. Reprod. Fertil.* suppl. 16, 39-59.
442a Nathanielsz, P.W., Comline, R.S. and Silver, M. (1973) Uterine activity following intravenous administration of oxytocin in the foetal sheep. *Nature* 243, 471-472.
443. Nathanielsz, P.W., Comline, R.S., Silver, M. and Thomas, A.L. (1973) Thyroid function in the foetal lamb during the last third of gestation and parturition. *J. Endocrinol.* 58, 535-546.
444. Nathanielsz, P.W., Comline, R.S., Silver, M. and Thomas, A.L. (1974) Thyroid function in the foetal calf. *J. Endocrinol.* 61, lxxi.
445. Nathanielsz, P.W., Silver, M. and Comline, R.S. (1973) Plasma triiodothyronine concentration in the foetal and newborn lamb. *J. Endocrinol.* 58, 683-684.
446. Nathanielsz, P.W. and Thomas, A.L. (1974) The release of thyrotrophin in response to thyrotrophin releasing hormone (TRH) in the pregnant ewe, lamb foetus and neonatal lamb. *J. Physiol.* 242, 108-109P.
447. Neumann, F., Elger, W. and Kramer, M. (1966) Development of a vagina in male rats by inhibiting androgen receptors with an anti-androgen during the critical phase of organogenesis. *Endocrinology* 78, 628-632.

448. Newton, W.H. (1935) 'Pseudo-parturition' in the mouse and the relation of the placenta to post-partum oestrus. *J. Physiol.* 84, 196-207.
449. Nichols, T., Nugent, C.A. and Tyler, F.H. (1965) Diurnal variation in suppression of adrenal function by glucocorticoids. *J. Clin. Endocrinol. Metab.* 25, 343-349.
450. Nicholson, J.L. and Altman, J. (1972) The effects of early hypo- and hyperthyroidism on the development of rat cerebellar cortex I. Cell proliferation and differentiation. *Brain Res.* 44, 13-23.
451. Nicoll, C.S. (1974) Physiological actions of prolactin. In: *Handbook of Physiology* Section 7, Vol. IV. The Pituitary Gland and its Neuroendocrine Control. Part 2, (Am. Physiol. Soc.) pp. 253-292.
452. Nikitovitch, M. and Knobil, E. (1955) Placental transfer of thyrotropic hormone in the rat. *J. Clin. Endocrinol. Metab.* 15, 837-838.
453. Noddle, B.A. (1964) Transfer of oxytocin from the maternal to the foetal circulation in the ewe. *Nature* 203, 414.
453a Novy, M.J., Cook, M.J. and Manaugh, L. (1974) Indomethacin block of normal onset of parturition in primates. *Am. J. Obstet. Gynecol.* 118, 412-416.
454. Novy, M.J., Piasecki, G. and Jackson, B.T. (1974) Effect of prostaglandins E_2 and $F_{2\alpha}$ on umbilical blood flow and fetal hemodynamics. *Prostaglandins* 5 (6) 543-555.
455. Oakey, R.E. (1970) The progressive increase in estrogen production in human pregnancy. An appraisal of the factors responsible. *Vitam. Horm.* 28, 1-36.
456. Obst, J.M. and Seamark, R.F. (1972) Plasma oestrogen concentrations in ewes during parturition. *J. Reprod. Fertil.* 28, 161-162.
457. O'Donohoe, N.W. and Holland, P.D.J. (1968) Familial congenital adrenal hypoplasia. *Arch. Dis. Child.* 43, 717-723.
458. Olley, P.M., Coceani, F. and Kent, G. (1974) Inactivation of prostaglandin E1 by the lungs of the foetal lamb. *Experientia* 30, 58-59.
459. Oppenheimer, J.H. and Surks, M.I. (1974) Quantitative aspects of hormone production, distribution, metabolism and activity. In: *Handbook of Physiology*. Section VII. Endocrinology III Thyroid. Eds.: Greer, M.A. and Solomon, D.H. (American Physiological Society) pp. 197-214.
459a Paisey, R.B. and Nathanielsz, P.W. (1971) Plasma cortisol levels in the newborn lamb from birth to 30 days. *J. Endocrinol.* 50, 701-702.
460. Pakravan, P., Kenny, F.M., Depp, R. and Allen, A.C. (1974) Familial congenital absence of adrenal glands, evaluation of glucocorticoid, mineralocorticoid and estrogen metabolism in the perinatal period. *J. Pediatr.* 84, 74-78.
461. Parker, L.N. and Noble, E.P. (1967) Prenatal glucocorticoid administration and the development of the epinephrine-forming enzyme. *Proc. Soc. Exp. Biol. Med.* 126, 734-737.
462. Pepe, G.J. and Townsley, J.D. (1975) Cortisol metabolism in the baboon during pregnancy and the post-partum period. *Endocrinology* 96, 587-591.
463. Perks, A.M. and Vizsolyi, E. (1973) Studies of the neurohypophysis in foetal mammals. In: *Foetal and Neonatal Physiology*. Proc. Sir Joseph Barcroft Centenary Symp. Eds.: Comline, R.S., Cross, K.W., Dawes, G.S. and Nathanielsz, P.W. (Cambridge University Press) pp. 430-438.
464. Peterson, N.A., Nataj, B.M., Chaikoff, I.L. and Ragupathy, E. (1966) Uptake of injected ^{131}I-labelled thyroxine, triiodothyronine and iodide by rat brain during various stages of development. *J. Neurochem.* 13, 933-943.

465. Peterson, R.R. and Young, W.C. (1952) The problem of placental permeability for thyrotropin, propylthiouracil and thyroxine in the guinea pig. *Endocrinology* 50, 218-225.
466. Phillippo, M., Lawrence, C.B. and Mellor, D.J. (1975) Changes of catecholamine concentrations in maternal and foetal plasma and in allantoic fluid at parturition in sheep and cows. *J. Endocrinol.* 65, 42-43P.
467. Pickering, D.L. (1964) Maternal thyroid hormone in the developing fetus. *Am. J. Dis. Child.* 107, 567-573.
468. Picon, R. (1971) Effects of the rat fetal testis on the regression of the Mullerian Ducts in vitro. In: *Hormones in Development*. Eds.: Hamburgh, M. and Barrington, E.J.W. (Appleton-Century-Crofts) pp. 645-650.
469. Pierrepoint, C.G., Anderson, A.B.M., Turnbull, A.C. and Griffiths, K. (1973) In vivo and in vitro studies of steroid metabolism by the sheep placenta. In: *The Endocrinology of Pregnancy and Parturition; Experimental Studies in the Sheep*. Ed.: Pierrepoint, C.G. (Tenovus Library Series, Alpha Omega Alpha, Cardiff) pp. 40-53.
470. Pietras, R.J. and Szego, C.M. (1975) Endometrial cell calcium and oestrogen action. *Nature* 253, 357-359.
471. Pinto, R.M., Leon, C., Mazzocco, N. and Scassera, V. (1967) Action of estradiol-17β at term and at onset of onset of labor. *Am. J. Obstet. Gynecol.* 98, 540-546.
472. Pittman, J.A., Brown, R.W. and Register, H.B. (1962) Biological activity of 3,3′,5′-triiodo-DL-thyronine. *Endocrinology* 70, 79-83.
473. Plaskett, L.G. (1962) Catabolism of doubly labelled thyroxine in vivo. *Nature* 195, 961-963.
474. Platzker, A.C.G., Kitterman, J.A., Mescher, E.J., Clements, J.A. and Tooley, W.H. (1975) Surfactant in the lung and tracheal fluid of the fetal lamb and acceleration of its appearance by dexamethasone. *Pediatrics* 56, 554-561.
475. Postel, S. (1957) Placental transfer of perchlorate and triiodothyronine in the guinea pig. *Endocrinology* 60, 53-66.
476. Pretell, G.A.F., Moncloa, R., Salinas, A., Kawano, R., Guerra-Garcia, L., Gutierrez, L., Beteta, L., Pretell, J. and Wan, M. (1969) Prophylaxis and treatment of endemic goitre in Peru with iodized oil. *J. Clin. Endocrinol. Metab.* 29, 1586-1593.
477. Pretell, E.A., Palacios, P., Tello, L., Wan, M., Utiger, R.D. and Stanbury, J.B. (1974) Iodine deficiency and the maternal-fetal relationship. In: *Endemic Goitre and Cretinism: Continuing Threats to World Health*. WHO Scientific Publication No. 292. Eds.: Dunn, J.T. and Medeiros-Neto, G.A. (WHO Washington, D.C.).
478. Pretell, E.A., Torres, T., Zenteno, V., Tello, L., Cornejo, M. (1972) Prophylaxis of endemic goitre with iodized oil in rural Peru. In: *Human Development and the Thyroid Gland: Relation to Endemic Cretinism*. Eds.: Stanbury, J.B. and Kroc, R.L. (Plenum Press, New York) pp. 249-288.
479. Pupkin, M.J., Schomberg, D.W., Nagey, D.A. and Crenshaw, C. (1975) Effect of exogenous dehydroepiandrosterone upon the fetoplacental biosynthesis of estrogens and its effect upon uterine blood flow in the term pregnant ewe. *Am. J. Obstet. Gynecol.* 121, 227-232.
480. Querido, A. and Swaab, D.F. (Eds.) (1975) *Brain Development and Thyroid Deficiency*. (North-Holland Publ. Co., Amsterdam-New York).
481. Rac, R., Hill, G.N. and Pain, R.W. (1968) Congenital goitre in Merino Sheep due

to an inherited defect in the biosynthesis of thyroid hormone. *Res. Vet. Sci.* 9, 209-223.

482. Raiti, S., Holzman, G.B., Scott, R.L. and Blizzard, R.M. (1967) Evidence for the placental transfer of triiodothyronine in human beings. *New Engl. J. Med.* 277, 456-459.
483. Rawlings, N.C. and Ward, W.R. (1973) Notes added in proof. In: *The Endocrinology of Pregnancy and Parturition; Experimental Studies in the Sheep*. Ed.: Pierrepoint, C.G. (Tenovus Library Series, Alpha Omega Alpha, Cardiff) pp. 171-173.
484. Rea, C. (1898) Prolonged gestation, acrania monstrosity and apparent placenta praevia on one obstetrical case. *J. Am. Med. Assoc.* 30, 1166-1167.
485. Rees, L.H., Chard, T., Evans, S.W. and Letchworth, A.T. (1975) Placental origin of ACTH in normal human pregnancy. *Nature* 254, 620-622.
486. Rees, L.H., Cook, D.M., Kendall, J.W., Allen, C.F., Kramer, R.M., Ratcliffe, J.G. and Knight, R.A. (1971) A radioimmunoassay for rat plasma ACTH. *Endocrinology* 89, 254-261.
487. Rees, L.H., Jack, P.M.B., Thomas, A.L. and Nathanielsz, P.W. (1975) Role of foetal adrenocorticotrophin during parturition in sheep. *Nature* 253, 274-275.
488. Reichlin, S. (1974) Regulation of somatotrophic hormone secretion. In: *Handbook of Physiology*, Section 7, Vol. IV. The Pituitary Gland and its Neuroendocrine Control, Part 2, Chapter 37. Eds.: Knobil, E. and Sawyer, W.H. (American Physiological Society) pp. 405-447.
489. Resko, J.A., Malley, A., Begley, D. and Hess, D.L. (1973) Radioimmunoassay of testosterone during fetal development of the rhesus monkey. *Endocrinology* 93, 156-161.
490. Resko, J.A., Ploem, J.G. and Stadelman, L. (1975) Estrogens in fetal and maternal plasma of the rhesus monkey. *Endocrinology* 97, 425-430.
491. Resnik, R., Battaglia, F.C., Makowski, E.L. and Meschia, G. (1975) The effect of actinomycin-D on estrogen-induced uterine blood flow. *Am. J. Obstet. Gynecol.* 122, 273-277.
492. Reynolds, J.W. and Mirkin, B.L. (1973) Urinary steroid levels in newborn infants with intrauterine growth retardation. *J. Clin. Endocrinol. Metab.* 36, 576-581.
493. Rieutort, M. (1972) Dosage radioimmunologique de l'Hormone somatotrope de Rat a l'aide d'une nouvelle technique de séparation. *CR Hebd. Séances Acad. Sci. (Paris)* 274, 3589-3592.
494. Roberts, E.M. (1966) The use of intravaginal sponges impregnated with 6-methyl-17 acetoxy progesterone (MAP) to synchronize ovarian activity in cyclic merino ewes. *Proc. Aust. Soc. Anim. Prod.* 6, 32-37.
495. Robertson, H.A. and Smeaton, T.C. (1973) The concentration of unconjugated oestrone, oestradiol-17α and oestradiol-17β in the maternal plasma of the pregnant ewe in relation to the initiation of parturition and lactation. *J. Reprod. Fertil.* 35, 461-468.
496. Robin, N.I., Fang, V.S., Selenkow, H.A., Piasecki, G.J., Rauschecker, H.F.J. and Jackson, B.T. (1970) Maternal-foetal thyroxine exchange in the pregnant sheep. *Clin. Res.* 18, 370.
497. Robin, N.I., Refetoff, S., Fang, V. and Selenkow, H.A. (1969) Parameters of thyroid function in maternal and cord serum at term pregnancy. *J. Clin. Endocrinol. Metab.* 29, 1276-1280.

498. Robinson, J.S. and Thorburn, G.D. (1974) The initiation of labour. *Br. J. Hosp. Med.* 12, 15-22.
499. Robinson, T.J. (1965) Use of progestogen-impregnated sponges inserted intravaginally or subcutaneously for the control of the oestrus cycle in the sheep. *Nature* 206, 39-41.
500. Roos, B.A. (1974) Effect of ACTH and cAMP on human adrenocortical growth and function in vitro. *Endocrinology* 94, 685-690.
501. Roos, T.B. (1967) Steroid synthesis in embryonic and foetal rat adrenal tissue. *Endocrinology* 81, 716-728.
502. Rosenfeld, C.R., Killam, A.P., Battaglia, F.C., Makowski, E.L. and Meschia, G. (1973) Effect of estradiol-17β on the magnitude and distribution of uterine blood flow in nonpregnant, oophorectomized ewes. *Pediatr. Res.* 7, 139-148.
503. Rossdale, P., Silver, M., Comline, R.S. and Nathanielsz, P.W. (1973) Plasma cortisol in the foal during the late fetal and early neonatal period. *Res. Vet. Sci.* 15, 395-397.
504. Rurak, D.W. (1976) Plasma vasopressin in foetal lambs. *J. Physiol.* 256, 36-37P.
505. Ryan, K.J. (1971) Endocrine control of gestational length. *Am. J. Obstet. Gynecol.* 109, 299-306.
506. Ryan, K.J. and Hopper, B.R. (1974) Placental biosynthesis and metabolism of steroid hormones in primates. *Contrib. Primatol.* 3, 258-283.
507. Sachs, H., Fawcett, P., Takabatake, Y. and Portanova, R. (1969) Biosynthesis and release of vasopressin and neurophysin. *Rec. Prog. Horm. Res.* 25, 447-491.
508. Salazar, H., MacAulay, M.A., Charles, D. and Pardo, M. (1969) The human hypophysis in anencephaly. 1. Ultrastructure of the pars distalis. *Arch. Pathol.* 87, 201-211.
509. Schofield, B.M. (1968) Parturition. In: *Advances in Reproductive Physiology*, Vol. III. Ed.: McLaren, A., pp. 9-32.
510. Schultz, F.M. and Wilson, J.D. (1974) Virilization of the Wolffian duct in the rat fetus by various androgens. *Endocrinology* 94, 979-986.
511. Schultz, M.A., Forsander, J.B., Chez, R.A. and Hutchinson, D.L. (1965) The bi-directional placental transfer of I^{131} 3,5,3'-triiodothyronine in the rhesus monkey. *Pediatrics* 35, 743-752.
512. Selye, H., Collip, J.B. and Thompson, D.L. (1935) Endocrine interrelations during pregnancy. *Endocrinology* 19, 151-159.
513. Seppala, M., Aho, I, Tissari, A. and Ruoslahti, E. (1972) Radioimmunoassay of oxytocin in amniotic fluid, fetal urine, and meconium during late pregnancy and delivery. *Am. J. Obstet. Gynecol.* 114, 788-795.
514. Serron-Ferre, M., Lawrence, C.C. and Jaffe, R.B. (1976) Control of cortisol secretion by the human fetal adrenal. *Gynecol. Invest.* (In press).
515. Shearman, R.P., Jools, N.D. and Smith, I.D. (1972) Maternal and fetal venous plasma steroids in relation to parturition. *J. Obstet. Gynaecol. Br. Commonw.* 79, 212-215.
516. Shelley, H.J. (1973) The use of chronically catheterized foetal lambs for the study of foetal metabolism. In: *Foetal and Neonatal Physiology*. Eds.: Comline, R.S., Cross, K.W., Dawes, G.S. and Nathanielsz, P.W. (Cambridge University Press) pp. 360-381.
517. Shephard, T.H. (1967) Onset of function in the human fetal thyroid: biochemical

and radioautographic studies from organ culture. *J. Clin. Endocrinol. Metab.* 27, 945-958.
518. Shephard, T.H. (1968) Development of the human fetal thyroid. *Gen. Comp. Endocrinol.* 10, 174-181.
519. Sheppard, H., Wiggan, G. and Tsien, W.H. (19) Structure–activity relationships for inhibitors in phosphodiesterase from erythrocytes and other tissues. *Adv. Cyclic Nucleotide Res.* 1, 103-112.
520. Sherwood, O.D., Rosentreter, K.R. and Birkhimer, M.L. (1975) Development of a radioimmunoassay for porcine relaxin using ^{125}I-labelled polytyrosyl–relaxin. *Endocrinology* 96, 1106-1113.
521. Short, R.V. (1974) Sexual differentiation of the brain of the sheep. In: *International Symposium on Sexual Endocrinology of the Perinatal Period.* (INSERM) Vol. 32, pp. 121-142.
522. Short, R.V., Smith, J., Mann, T., Evans, E.P., Hallet, J., Fryer, A. and Hamerton, J.L. (1969) Sexual differentiation of the brain of the sheep. *Cytogenetics* 8, 369-388.
523. Siiteri, P.K. and MacDonald, P.C. (1966) Placental estrogen biosynthesis during human pregnancy. *J. Clin. Endocrinol. Metab.* 26, 751-761.
524. Silver, M. (1976) Fetal energy metabolism. In: *Fetal Physiology and Medicine.* Eds.: Beard, R.W. and Nathanielsz, P.W. (W.B. Saunders).
525. Silver, M., Steven, D.H. and Comline, R.S. (1973) Placental exchange and morphology in ruminants and the mare. In: *Foetal and Neonatal Physiology.* Eds.: Comline, R.S., Cross, K.W., Dawes, G.S. and Nathanielsz, P.W. (Cambridge University Press) pp. 245-271.
526. Skowsky, W.R., Bashore, R.A., Smith, F.G. and Fisher, D.A. (1973) Vasopressin metabolism in the foetus and newborn. In: *Foetal and Neonatal Physiology.* Proc. Sir Joseph Barcroft Centenary Symp. Eds.: Comline, R.S., Cross, K.W., Dawes, G.S. and Nathanielsz, P.W. (Tenovus Library Series, Alpha Omega Alpha, Cardiff) pp. 439-447.
527. Slob, A.K., Goy, R.W. and Van der Werff ten Bosch, J.J. (1973) Sex differences in growth of guinea pigs and their modification by neonatal gonadectomy and prenatally administered androgen. *J. Endocrinol.* 58, 11-19.
528. Small, C.W. and Watkins, W.B. (1975) Oxytocinase-immunohistochemical demonstration in the immature and term human placenta. *Cell Tissue Res.* 162, 531-539.
529. Smith, G.C. (1970) Ultrastructural studies on the median eminence of neonatal rats. *J. Anat.* 106, 200.
530. Smith, D.W., Blizzard, R.M. and Wilkins, L. (1957) The mental prognosis in hypothyroidism of infancy and childhood: a review of 128 cases. *Pediatrics* 19, 1011-1022.
531. Smith, I.D. and Shearman, R.P. (1974) Fetal plasma steroids in relation to parturition. *J. Obstet. Gynaecol.* 81, 11-15.
532. Solomon, S., Bird, C.E., Ling, W., Iwamiya, M. and Young, P.C.M. (1967) Formation and metabolism of steroids in the fetus and placenta. *Rec. Prog. Horm. Res.* 23, 297-335.
533. Starling, M.B. and Elliott, R.B. (1974) The effects of prostaglandins, prostaglandin inhibitors and oxygen on the closure of the ductus arteriosus, pulmonary arteries and umbilical vessels in vitro. *Prostaglandins* 8, 187-203.

534. Steele, P.A., Flint, A.P.F. and Turnbull, A.C. (1975) Evidence of C17-20 lyase activity in ovine foeto–placental tissue. *J. Endocrinol.* 64, 41P.
535. Sterling, K. (1970) Significance of circulating triiodothyronine. *Rec. Prog. Horm. Res.* 26, 249-286.
536. Sterling, K. (1974) The nature of iodine in plasma. In: *Handbook of Physiology*, Section VII, Endocrinology III. Thyroid. Eds.: Greer, M.A. and Solomon, D.H. (Am. Physiol. Soc., Washington, D.C.) pp. 179-186.
537. Steven, D.H. (1975) Separation of the placenta in the ewe: an ultrastructural study. *Quart. J. Exp. Physiol.* 60, 37-44.
538. Stokes, H. and Boda, J.M. (1968) Immunofluorescent localisation of growth hormone and prolactin in the adenohypophysis of foetal sheep. *Endocrinology* 83, 1362-1366.
539. Strott, C.A., Sundel, H. and Stahlman, M.T. (1974) Maternal and fetal plasma progesterone, cortisol, testosterone and 17β-estradiol in preparturient sheep: response to fetal ACTH infusion. *Endocrinology* 95, 1327-1339.
539a Swaab, D.F., Boer, K. and Honnebier, W.J. (1977) The fetal hypothalamus and pituitary in the onset and the course of parturition. In: *The Fetus and Birth*. Ed.: Maeve O'Connor. CIBA Foundation Symposium No. 47, New Series. (Elsevier, Amsterdam).
540. Swaab, D.F. and Honnebier, W.J. (1973) The influence of removal of the fetal rat brain upon intra-uterine growth of the fetus and the placenta and on gestation length. *J. Obstet. Gynaecol. Br. Commonw.* 80, 589-597.
541. Swaab, D.F. and Honnebier, W.J. (1974) The role of the fetal hypothalamus in development of the feto-placental unit and in parturition. In: *Integrative Hypothalamic Activity*. Eds.: Swaab, D.F. and Schadé, J.P. *Prog. Brain Res.* Vol. 41, (Elsevier, Amsterdam) pp. 255-280.
542. Swanson, H.E. and Werff ten Bosch, J.J. Van Der (1963) Sex differences in growth of rats and their modification by a single injection of testosterone proprionate shortly after birth. *J. Endocrinol.* 26, 197-207.
543. Symonds, E.M. and Furber, I. (1973) Plasma renin levels in the normal and anephric fetus at birth. *Biol. Neonate* 23, 133-138.
544. Talbert, L.M., Easterling, W.E. and Potter, H.D. (1973) Maternal and fetal plasma levels of adrenal corticoids in spontaneous vaginal delivery and cesarian section. *Am. J. Obstet. Gynecol.* 117, 554-559.
545. Tamby Raja, R.L., Anderson, A.B.M. and Turnbull, A.C. (1974) Endocrine changes in premature labour. *Br. Med. J.* 4, 67-71.
546. Tata, J.R. (1974) Growth and developmental action of thyroid hormones at the cellular level. In: *Handbook of Physiology*, Section 7, Endocrinology III. Thyroid. Eds.: Greer, M.A. and Solomon, D.H. (Am. Physiol. Soc., Washington, D.C.) pp. 469-478.
547. Tato, L., Marc, V.L., Prevot, C. and Rappaport, R. (1975) Early variations of plasma somatomedin activity in the newborn. *J. Clin. Endocrinol. Metab.* 40, 534-536.
548. Taurog, A., Tong, W. and Chaikoff, I.L. (1958) Thyroid ^{131}I metabolism in the absence of the pituitary: the untreated hypophysectomised rat. *Endocrinology* 62, 646-663.
549. Taurog, A., Tong, W. and Chaikoff, I.L. (1958) Thyroid ^{131}I metabolism in the

absence of the pituitary: the hypophysectomised rat treated with thyrotropic hormone. *Endocrinology* 62, 664-676.
550. Teoh, E.S., Spellacy, W.N. and Buhi, W.C. (1971) Human chorionic somatomammotrophin (HCS): a new index of placental function. *J. Obstet. Gynecol. Br. Commonw.* 78, 673-685.
551. Thomas, A.L., Jack, P.M.B., Manns, J.G. and Nathanielsz, P.W. (1975) Effect of synthetic thyrotrophin releasing hormone on thyrotrophin and prolactin concentrations in the peripheral plasma of the pregnant ewe, lamb fetus and neonatal lamb. *Biol. Neonate* 26, 109-116.
552. Thorburn, G.D. (1974) The role of the thyroid gland and kidneys in fetal growth. In: *Size at Birth*. CIBA Foundation Symp. 27, 185-214.
553. Thorburn, G.D., Cox, R.I., Currie, W.B., Restall, B.J. and Schneider, W. (1973) Prostaglandin F and progesterone concentrations in the utero-ovarian venous plasma of the ewe during the oestrous cycle and early pregnancy. *J. Reprod. Fertil.* suppl. 18, 151-158.
554. Thorburn, G.D. and Hopkins, P.S. (1973) Thyroid function in the foetal lamb. In: *Foetal and Neonatal Physiology*. Proc. Sir Joseph Barcroft Centenary Symp. Eds.: Comline, R.S., Cross, K.W., Dawes, G.S. and Nathanielsz, P.W. (Cambridge University Press) pp. 488-507.
554a Thorburn, G.D., Nicol, D.H., Bassett, J.M., Shutt, D.A. and Cox, R.I. (1972) Parturition in the goat and sheep: changes in corticosteroids, progesterone, oestrogens and prostaglandin F. *J. Reprod. Fertil.* suppl. 16, 61-84.
555. Tong, W. (1974) Actions of thyroid-stimulating hormone. In: *Handbook of Physiology*. Section VII. Endocrinology, Volume III Thyroid. Eds.: Greer, M.A. and Solomon, D.H. (Am. Physiol. Soc. Washington, D.C.) pp. 255-284.
556. Towell, M.E. and Liggins, G.C. (1976) The effect of labour on uterine blood flow in the pregnant ewe. *Quart. J. Exp. Physiol.* 61, 23-33.
557. Tuppy, H. (1968) The influence of enzymes on neurohypophysial hormones and similar peptides. In: *Handbook of Experimental Pharmacology*. Ed.: Berde, B. (Springer, Berlin) pp. 67-129.
557a Turkington, R.W., Frantz, W.L. and Majumder, G.C. (1973) Effector–receptor relations in the action of prolactin. In: *Human Prolactin: Proc. Int. Symp. Human Prolactin*, Brussels, June 12-14, 1973. Eds.: Pasteels, J.L. and Robyn, C., pp. 25-34.
558. Turnbull, A.C. and Anderson, A.B.M. (1969) The influence of the foetus on myometrial contractility. In: *Progesterone: The Regulatory Effect on the Myometrium*. Eds.: Wolstenholme, G.E.W. and Knight, J. (Churchill) pp. 106-113.
559. Turnbull, A.C., Patten, P.T., Flint, A.P.F., Kierse, M.J.N.C., Jeremy, J.Y. and Anderson, A.B.M. (1974) Significant fall in progesterone and rise in oestradiol levels in human peripheral plasma before onset of labour. *Lancet* 1, 101-104.
560. Tyson, J.E., Hwang, P., Guyda, H. and Friesen, H.G. (1972) Studies of prolactin secretion in human pregnancy. *Am. J. Obstet. Gynecol.* 113, 14-20.
561. Urquhart, J. (1974) Physiological actions of adrenocorticotropic hormone. In: *Handbook of Physiology*, Section 7, Endocrinology, Volume IV. The pituitary gland and its neuroendocrine control Part 2. Eds.: Knobil, E. and Swayer, W.H. (Am. Physiol. Soc., Washington, D.C.) pp. 133-158.
562. Van Assche, F.A., Gepts, W. and de Gasparo, M. (1969) The endocrine pancreas in anencephalics. *Horm. Metab. Res.* 1, 251-252.

563. Van Leusden, H.A. and Villee, C.A. (1966) Formation of estrogens by hydatidiform moles in vitro. *J. Clin. Endocrinol. Metab.* 26, 842-846.
564. Van Petten, G.R. (1975) Pharmacology and the fetus. *Br. Med. Bull.* 31 (1) 75-79.
565. Van Wagenen, G. and Newton, W.H. (1943) Pregnancy in the monkey after removal of the fetus. *Surg. Gynecol. Obstet.* 77, 539-543.
566. Van Wagenen, G. and Simpson, M.E. (1954) Testicular development in the rhesus monkey. *Anat. Rec.* 118, 231-252.
567. Van Wyk, J.J., Underwood, L.E., Hintz, R.L., Clemmons, D.R., Voina, S.J. and Weaver, R.P. (1974) The somatomedins: a family of insulin like hormones under growth hormone control. *Rec. Prog. Horm. Res.* 30, 259-294.
568. Van Wynsberghe, D.M. and Klitgaard, H.M. (1973) The effects of thyroxine and triiodothyroacetic acid on neonatal development in the rat. *Biol. Neonate* 22, 444-450.
569. Vane, J.R. and Williams, K.I. (1973) The contribution of prostaglandin production to contractions of the isolated uterus of the rat. *Br. J. Pharmacol.* 48, 629-639.
570. Venning, E.H., Dyrenfurth, I., Lowenstein, L. and Beck, J.C. (1959) Metabolic studies in pregnancy and the puerperium. *J. Clin. Endocrinol. Metab.* 19, 403.
570a Verney, E.B. (1947) The antidiuretic hormone and the factors which determine its release. *Proc. R. Soc.* (Ser. B) 135, 25-106.
571. Vizsolyi, E. and Perks, A.M. (1969) New neurohypophyseal principles in foetal mammals. *Nature* 223, 1169-1171.
572. Vizsolyi, E. and Perks, A.M. (1974) The effect of arginine vasotocin on the isolated amniotic membrane of the guinea-pig. *Can. J. Zool.* 52, 371-386.
573. Wagner, W.C., Thompson, F.N., Evans, L.E. and Molokwu, E.C.I. (1974) Hormonal mechanisms controlling parturition. *J. Anim. Sci.* 38, suppl. 1, 39-54.
574. Wakerley, J.B. and Lincoln, D.W. (1973) The milk-ejection reflex of the rat: a 20- to 40-fold acceleration in the firing of paraventricular neurones during the release of oxytocin. *J. Endocrinol.* 57, 477-493.
575. Wallace, A.L.C. and Bassett, J.M. (1970) Plasma growth hormone concentrations in sheep measured by radioimmunoassay. *J. Endocrinol.* 47, 21-36.
576. Wallace, A.L.C., Stacy, B.D. and Thorburn, G.D. (1972) The fate of radioiodinated sheep growth hormone in intact and nephrectomized sheep. *Pflügers Arch. Eur. J. Physiol.* 331, 25-37.
577. Wallace, A.L.C., Stacy, B.D. and Thorburn, G.D. (1973) Regulation of growth hormone secretion in the ovine foetus. *J. Endocrinol.* 58, 89-95.
578. Weidemann, E. and Schwartz, E. (1972) Suppression of growth hormone-dependent human sulfation factor by estrogen. *J. Clin. Endocrinol. Metab.* 34, 51-58.
579. Welch, R.A.S., Frost, D.L. and Bergman, M. (1973) The effect of administering ACTH directly to the foetal calf. *NZ Med. J.* 78, 365.
580. Welsch, C.W., Squiers, M.D., Cassell, E., Chen, C.L. and Meites, J. (1971) Median eminence lesions and serum prolactin: influence of ovariectomy and ergocornine. *Am. J. Physiol.* 221, 1714-1717.
581. Widnell, C.C. and Tata, J.R. (1966) Additive effects of thyroid hormone, growth hormone and testosterone on deoxyribonucleic acid-dependent ribonucleic acid polymerase in rat-liver nuclei. *Biochem. J.* 98, 621-628.
582. Wilhelmi, A.E. (1974) Chemistry of growth hormone. In: *Handbook of Physiology*, Section 7, Volume IV. The pituitary gland and its neuroendocrine control. Part 2.

Eds.: Knobil, E. and Swayer, W.H. (Am. Physiol. Soc., Washington, D.C.) pp. 59-78.
583. Wilson, D.W., Pierrepoint, C.G. and Griffiths, K. (1973) The radioimmunoassay of ovine prolactin during pregnancy. Biochemical Society Trans. 1, 172-175.
584. Wilson, J.D. (1974) Metabolism of testicular androgens. In: *Handbook of Physiology*, Section 7. Endocrinology (Am. Physiol. Soc., Washington, D.W.) Chapter 25.
585. Warren, D.W., Haltmeyer, G.C. and Eik-Nes, K.B. (1973) Testosterone in the fetal rat testis. *Biol. Reprod.* 8, 560-565.
586. Witschi, E. (1956) *Development of Vertebrates*. (Saunders.)
587. Wilson, J.D. (1973) Testosterone uptake by the urogenital tract of the rabbit embryo. *Endocrinology* 92, 1192-1199.
588. Wilson, J.D. and Siiteri, P.K. (1973) Developmental pattern of testosterone synthesis in the fetal gonad of the rabbit. *Endocrinology* 92, 1182-1191.
589. Wilson, L., Cenedella, R.J., Butcher, R.L. and Inskeep, E.K. (1971) Progesterone treatment on ovine endometrial prostaglandins. *J. Anim. Sci.* 33, 273.
590. Winters, A.J., Oliver, C., Colston, C., MacDonald, P.C. and Porter, J.C. (1974) Plasma ACTH levels in the human fetus and neonate as related to age and parturition. *J. Clin. Endocrinol. Metab.* 39, 269-273.
591. Wintour, M.E., Brown, E.H., Hardy, A.K., McDougall, J.G., Oddie, C.J. and Whipp, G.T. (1974) Ontogeny of secretion and control of ovine foetal adrenal cortex. *J. Steroid Biochem.* 5, 366.
592. Wintour, E.M., Brown, E.H., Denton, D.A., Hardy, K.J., McDougall, J.G., Oddie, C.J. and Whipp, G.T. (1975) The ontogeny and regulation of corticosteroid secretion by the ovine foetal adrenal. *Acta Endocrinol.* 79, 301-316.
593. Woeber, K.A. and Ingbar, S.H. (1974) Interactions of thyroid hormones with binding proteins. In: *Handbook of Physiology*. Section VII. Endocrinology III, Thyroid. Eds.: Greer, M.A. and Solomon, D.H. (Am. Physiol. Soc., Washington, D.C.) pp. 187-196.
594. Wong, M.S.F., Cox, R.I., Currie, W.B. and Thorburn, G.D. (1972) Changes in oestrogen sulphoconjugates in the foetal plasma of sheep and goats during late gestation. In: *Proc. 15th Annu. Meet. Endocrine Soc. Aust.* Sydney, August 1972. Vol. 15, Abstract 8.
595. Wu, B., Kikkawa, Y., Orzalesi, M.M., Motoyama, E.K., Kaibara, M., Zigas, C.J. and Cook, C.D. (1973) The effect of thyroxine on the maturation of fetal rabbit lungs. *Biol. Neonate* 22, 161-168.
596. Yamazaki, E., Noguchi, A. and Slingerland, D.W. (1960) The in vitro metabolism of iodide, iodotyrosine and thyroxine by human placenta. Journal of Clinical Endocrinol 20, 794-797.
597. Yates, F.E. and Marran, J.W. (1974) Stimulation and inhibition of adrenocorticotrophin release. In: *Handbook of Physiology*, Section 7. Endocrinology, Volume IV. The Pituitary Gland and its Neuroendocrine Control, Part 2 (Am. Physiol. Soc., Washington, D.C.) pp. 367-404.
598. Yates, F.E., Russell, S.M., Dallman, M.F., Hedge, G.A., McCann, S.M. and Dhariwal, A.P.S. (1971) Potentiation by vasopressin of corticotrophin release induced by corticotrophin-releasing factor. *Endocrinology* 88, 3-15.
599. Young, F.G. (1940) The pituitary gland and carbohydrate metabolism. *Endocrinology* 26, 345-351.

600. Zarrow, M.X., Philpott, J.E. and Denenberg, V.H. (1970) Passage of ^{14}C-4-corticosterone from rat mother to foetus and neonate. *Nature* 226, 1058-1059.
601. Zimmerman, E.A., Carmel, P.W., Husain, M.K., Ferin, M., Tannenbaum, M., Frantz, A.G. and Robinson, A.G. (1973) Vasopressin and neurophysin: high concentrations in monkey hypophyseal portal blood. *Science* 182, 925-927.

Subject index

ACTH, 10, 17–19, 94, 96, 103, 125–150, 152–155, 165, 171, 173, 174, 201
  adrenal sensitivity to, 141–146
  concentration in human fetus, 211
  fetal plasma concentrations in acute experiments, 127
  measurement of plasma concentrations, 125–131
  mode of action, 152
  phasic changes in secretion, 133–135, 137
  secretion and parturition, 196–199
  secretion rate, calculation, 135, 136
  secretion response to delivery stress, 129
  secretion response to surgery stress, 25
Adenohypophysis, 57, 59, 107, 152
Adenylate cyclase,
  activity in adrenal, 142, 143, 152
  stimulation by AVP, 91
Adrenal, 151–175
  axis, 17
    feedback mechanisms, 138–141
    response to stress, 148–150
  blood flow, 148, 197
  control of growth, 153
  corticosterone secretion by rat, 155
  effect of growth on cortisol production, 131–133
  human, 170
  in vitro sensitivity to ACTH, 18
  pituitary control of, 173–175
  possible role for inhibitors, 154
  primate cortex, fetal and definitive zones, 170
  role in parturition, 195, 201, 211–214
  secretion of cortisol, 136–138
  sensitivity,
    effect of dexamethasone infusions, 144
    effect on delivery post-surgery, 25
    to ACTH, 141, 154
    sub-human primate, 170
Adrenal–placental interaction in sheep and primate, 213
Adrenalectomy (fetal), effect on delivery in monkey, 210
Adrenalin, 14, 111, 113, 163
Aldosterone, 18, 105, 158
Allantoic fluid composition, 104
Amniotic fluid,
  composition, 104
  cortisol concentration, 171, 212
Anaesthesia, 24
Androgens,
  biosynthetic pathways, 32–34
  role in development, 34–37
Androstenedione, 180
Antidiuretic hormone, 18
Arachidonic acid, 205, 206
Arginine vasopression (AVP), 89–94, 99–106
  action as a CRF, 149
  fetal and maternal plasma concentrations, 99
  physiological role in the fetus, 100
  structure, 90
Arginine vasotocin (AVT), 91
  structure, 90
Autonomic nervous system, effects on thyroid hormones, 66

Barr body, 38
Blood gases, use in fetal monitoring, 23, 25, 26
Blood volume, 127

Body fluids,
 hormonal control, 103
 ionic changes post-operatively, 25
Bone differentiation, effect of thyroid hormones, 84, 85
Brain, 2
 sex differentiation, 43
 differentiation,
  effect of glucocorticoids, 133
  effect of thyroid hormones, 51, 85

Catecholamines, 112, 113
Catheterised fetal sheep preparation, 22–28
Cervical dilatation,
 oxytocin release, 206
 role of prostaglandins, 206, 214
Cervical dystocia, 204
Choline phosphotransferase system, 166–168
Chorionic gonadotropin, 107, 108
Chromosomes,
 autosomes, 38, 39
 sex chromosomes, 38, 39
Circadian rhythms,
 fetal breathing movements, 134
 in hormone secretion, 51, 125, 133–135
Congenital abnormalities,
 delayed parturition in the human, 210
 effect on parturition, 193
 effects of steroids, 169
Control mechanisms, 1
 anatomically addressed, 1
 chemically addressed, 1
 feedback loops, 5, 138–141
 hierarchical control, 6
 inductors, 5
Corpus luteum, 108
 progesterone production, 202
Corticosterone, 133, 149, 151–175
Corticotropin-like intermediate lobe peptide (CLIP), structure, 126
Corticotropin-releasing factor (CRF), 45, 105, 106, 147, 149
Cortisol, 18, 128–146, 149–175
 -binding proteins, 160, 161
 concentration in fetus, 128, 129
 infusion into fetus, effect on thyroid hormones, 70
 placental transfer, 173

 secretion,
  and parturition, 196–202, 211–214
  control by factors other than ACTH, 136–138
  effect of adrenal growth on, 131–133
  factors affecting prenatal changes, 130
Cortisone, 151
Cretinism, role of iodothyronines, 79–81
Cyclic AMP, 11, 91, 198, 207
Cyclic GMP, 11, 207
Cyclic nucleotides, role in parturition, 207
Cystine aminopeptidase (CAP), 91

Dehydroepiandrosterone (DHEA), 179–180, 185–187
Dexamethasone, 106, 144, 146, 150, 165, 166, 169, 181–184, 200–203, 212, 213
Differentiation, effect of thyroid hormones, 83–87
Dihydrotestosterone, 36, 41–43

Electroencephalograph (EEG),
 activity and fetal breathing movements, 134
 activity during gestation, 115
 post-operative changes, 25
 relation to GH secretion, 113
Encephalectomy, use for investigating adrenal function, 146
Endostyle, 53
Enzymes,
 ATPase, 61
 11β-dehydrogenase, 173
 17α-hydroxylase, 33, 153, 170, 178, 183, 186, 212
 11μ-hydroxylase, 153
 3μ-hydroxysteroid dehydrogenase, 33, 170, 186, 212
 Δ^5-isomerase, 33, 170
 17,20-lyase, 33, 170, 178, 186, 212
 phosphoethanolamine-N-methyltransferase, 14, 169
 phospholipase A, 205
 placental aromatase, 170, 180, 186
 primate liver 16α-hydroxylase, 185, 186
 5α-reductase, 33, 36, 41
 succinate dehydrogenase, 87
 sulphatase, 181

sulphokinase, 181
Epidural anaesthesia, 24
Estrogens, 32, 177–190
  effect on somatomedin, 117
  maternal plasma concentrations in primate, 188
  PG synthesis, 204
  placental permeability to, 182
  role in parturition, 204, 205, 214

Fetal autonomy, 5
Fetal membranes, 3
  arrangement in sheep, 104
  PG metabolism 214
Fetectomy, effect on progress of pregnancy, 208–210
Fetus, definition, 21

Gene expression, 10
Glucocorticoids, 113, 114, 128–146, 159–175, 178, 196–201
  administration during pregnancy, 212–213
  biosynthetic pathways, 151
  blood production rates, 160
  concentration changes after birth, 156
  concentration changes before birth, 161
  enzyme induction, 14
  induction of parturition, 201
  mechanisms of action in fetal sheep, 163–169
  negative feedback on ACTH secretion, 149
  plasma binding, 160
  secretion in late gestation, 156
Glucose concentration, effect on GH, 112
Gonadal development, 34–44
Growth hormone (GH), 101, 103, 104–118
  actions in fetus, 109
  concentration in fetus, 134
  control of secretion, 112
  effect of pituitary stalk section, 49
  fetal MCR and BPR, 108
  metabolism by kidney, 111
  pituitary and plasma concentrations, 108
  relation to cortisol, 114
  relation to SM, 111

Haemorrhage,
  calculation of fetal blood volume, 127
  fetal response, 101–103
Haemorrhagic stress,
  relation to ACTH release, 126, 127
  relation to measurement of AVP, 90
Heterochromatin, 38
Hormone, definition, 1
Hypophysectomy, 83
  effect on gestation, 193, 202
  effect on plasma ACTH concentration, 138
Hypothalamus, 57–59, 115, 146–150
Hypothalamo–hypophysial–portal system, 45–52
Hypoxia,
  release of ACTH, 101, 127, 148
  release of AVP, 93, 100, 101
  release of catecholamines, 101
  release of GH, 101

Indian ink, perfusion with, 45, 46
Indomethacin, 214
Insulin, growth-promoting action, 111
Iodothyronines, 54, 57–82
  placental permeability, 79
  placental transport, 75, 76
  role in cretinism, 79–81

Lecithin (lung surfactant), 84, 165–169
Lipotropic hormone (LPH), structure, 126
Lung differentiation, effect of thyroid hormones, 84
Luteinising hormone (LH), 33, 34, 51
Luteinising hormone-releasing hormone (LH-RH), 33, 34, 51
Lysosomes, role in PGF production, 205

Melanocyte-stimulating hormone (MSH), 10, 126, 152
Membranes, 3
  fetal, arrangement in sheep, 104
  PG synthesis, 189, 214
Mesonephric duct, 35
Methodology, 13–29
Metopirone, effects on fetal adrenal axis, 138, 139
Mullerian duct, 35–37, 41
Mullerian duct inhibitory factor, 35, 41

Nervous system, differentiation, effect of thyroid hormones, 85–87
Neurohypophysis, 89–106
Neurophysins, 92
Neurosecretion, 6, 7
Nyctohemeral rhythms, hormone secretion, 51, 125

Oxytocin, 18, 89–100
    bioassay, 92, 93
    destruction by CAP, 91
    effect of fetal injection, 96
    placental transport, 96–98
    physiological actions, 90
    release at parturition, 206, 207
    structure, 90

Parathormone, 6
Paraventricular nucleus, 89, 93
Parturition, 191, 216
    fetal signal, 196
Pituitary stalk section, 45
    effect on adrenal weight, 147
    effect on hormone secretion, 49–51
Placenta,
    cortisol to cortisone conversion, 4
    feto–placental unit, 3
    iodide trapping by the placenta, 66
    materno–placental unit, 4
    permeability to,
        AVP, 99
        cortisol, 159, 160, 173
        estrogens, 182
        hormones, 4
        iodothyronines, 79
        prolactin, 119
    progesterone production, 202
    structure, 3
    transport of iodothyronines,
        guinea pig, 75
        rat, 75
    transport of oxytocin, 94, 96
Placental lactogen (PL), 91, 107, 121–123
    concentration in sheep placenta, 121
    plasma and amniotic fluid concentrations, 121
Portal vessels, Indian ink perfusion studies, 46

Positive feedback, mechanisms in parturition, 206
Pre-operative preparation, 24
Primate model, 28
Progesterone, 182–190, 202–204, 213–214
    action on hypothalamic neurones, 204
    mechanism of prepartum fall in plasma concentration, 183
    umbilical arterio–venous differences, 188
    use pre-operatively, 24
Prolactin, 107, 118–121
    action on other hormone systems, 120
    amniotic fluid concentrations, 118
    pituitary and plasma concentrations, 118
    possible effects of pituitary stalk section, 119
    secretion after TRH challenge test, 119
Prostaglandins (PGs)
    effect of estrogens, 204
    effect on neurohypophysis, 206
    in fetal primate, 189–190
    in fetal sheep, 184–186
    in parturition, 96, 205–207, 214
    metabolites, 214
    structure, 205

Rathke's pouch, 13, 14, 89
Receptors,
    progesterone and estrogen binding, 203
    protein hormone binding, 12
    steroid binding, 11
Relaxin, 204
REM sleep, breathing movements, 134
Renin–angiotensin system, 93, 101, 102
Respiratory distress syndrome, 169
Respiratory movements, post-operative changes, 25

Sex-hormone-binding globulin, 188
Skin, differentiation, effect of thyroid hormones, 85
Somatomedin (SM), 68, 101
    changes in tissue responsiveness, 10
    influence of estrogens, 117
    relationship to GH, 111, 112, 116, 117
Steroids, androgenic, synthesis, 32, 33, 170, 171
Sulpho-conjugates,

of estrogens, 82, 179–182, 185–187
of T_3, 89, 93
Supra-optic nucleus (SON), 89, 93
Surfactant (dipalmitoyl lecithin), 165–169
Surgical techniques,
   encephalectomy, 47
   fetal decapitation, 47
   fetal vascular catheterisation, 22
   pituitary stalk section, 45, 49
Syncytiotrophoblast, 4
   production of hPL (hCS) and hCG, 107, 121

Techniques of investigation, 17
Testis, 31–44
   cortexenes and medullarenes, 39
   differentiation, 31, 32, 38, 39
Testosterone, 11, 12, 44, 180
   plasma concentrations, 187
Thyrocalcitonin, 55
Thyroid axis, 17
   development in rat, 73–75
   negative feedback in fetal rat, 74
Thyroid gland, 53–87
   development,
      human and sub-human primate, 83–87
      rat, 73–75
      ruminant, 53–57
   function in ruminant, 61–72
   sub-human primate studies, 81
Thyroid hormones, 53–87
   effect on differentiation, 83–87
   effects of pituitary stalk section, 63
   mode of action, 60
   placental permeability, 63
   structure–activity relationships, 59
Thyroidectomy,
   effect on fetal: placental weight ratio in primate fetus, 82
   effect on growth, 83
   effect on mental performance, 85
Thyrotropin (TSH), 18, 48, 49, 51, 57, 58, 62–68, 72–80, 86, 119
Thyrotropin-releasing hormone (TRH), 50, 51, 57, 73–75, 119
Thyroxine (T_4), 15, 16, 18, 51–79, 81, 82, 214
   biosynthesis, 55–57
   peripheral deiodination, 67
Transport of hormones in blood, 16
Triiodothyronine (T_3), 18, 57–82, 86, 214
   neonatal changes, 66
   plasma concentrations in fetus, 64
   reverse T_3, 58, 60–63, 67, 68, 73, 214
Twin pregnancies, effect of ACTH infusion, 141

Ultimobranchial body, 55
Urogenital sinus, 35–37, 41
Urogenital swelling, 35–37
Urogenital tubercle, 35–37, 41
Uterine contraction, 191, 192, 205–207

Vagosympathetic nerve trunk, effect of section on AVP secretion, 101
Vasopressin bioassay, 92
Vasotocin, 10
Veratrum californicum, 193, 210

Wolffian duct, 35–37, 41–43